29 jour
travers parafine — vide — 1 jour expos

Vibent   2000 —g   app(1, 3ᵉ

30 jour

tube bouché paraffine et tran   1 heu
dans flacon bouché émeri —   app (0)
dedans   2 Rap   9
dehors   2 "   /3

golas   Vide   golas même appareil vid
dehors   800 — 20
dedans   800   24

mouvement propre   800 —   24.

a servi a mesure corps actif
approximant   pas de fuites sérieuses   app. ordinair

paraffine   1 jour

avec jauge.   pas de fuites sérieuses

ap. nº 1   dedans   800   14"
ap. nº 1   dehors   800   23"

mouv. propre —   rien

impossible mesu. petites activité

THE
ELEMENTS

Pl. 13.

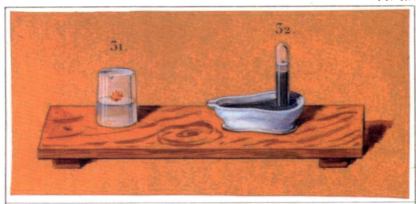

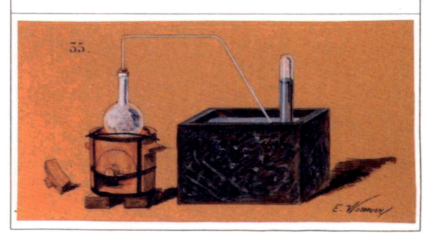

THE | ELEMENTS

A VISUAL HISTORY OF THEIR DISCOVERY

PHILIP BALL

THE UNIVERSITY OF CHICAGO PRESS

The University of Chicago Press, Chicago 60637

Published 2021

Printed in Singapore

30 29 28 27 26 25 24 23 22 21   1 2 3 4 5

ISBN-13: 978-0-226-77595-1 (cloth)
ISBN-13: 978-0-226-77600-2 (e-book)

DOI: https://doi.org/10.7208/chicago/9780226776002.001.0001

Conceived, designed and produced by
Quarto Publishing plc
The Old Brewery
6 Blundell Street
London N7 9BH
www.quartoknows.com

QUAR.336392

Library of Congress Cataloging-in-Publication Data

Names: Ball, Philip, 1962– author.
Title: The elements : a visual history of their discovery / Philip Ball.
Description: Chicago : University of Chicago Press, 2021. | Includes bibliographical references and index.
Identifiers: LCCN 2021005240 | ISBN 9780226775951 (cloth) | ISBN 9780226776002 (ebook)
Subjects: LCSH: Chemical elements—History.
Classification: LCC QD466 .B26 2021 | DDC 546.09—dc23
LC record available at https://lccn.loc.gov/2021005240

FRONT COVER: Antique illustration of iron combustion in pure oxygen. Artist unknown, first published in *L'Eau* by Gaston Tissandier, Hachette, Paris, 1873.

PREVIOUS PAGE: The action of water on potassium and ammonia, and preparation of ammonia, from J. Pelouze et E. Fremy, *Notions générales de chimie*, Paris: Victor Masson, 1853, plate XIII. National Central Library of Florence.

CONTENTS

INTRODUCTION

Among the many discoveries that humans have made about the world we live in, one of the most profound, as well as the most useful, is what it is made from. Every substance we can see and feel is composed of atoms—too small to see even with conventional light microscopes—of which there are just 90 or so varieties. What's more, many of those varieties are extremely rare; the familiar world encompasses perhaps just twenty to thirty of them. These varieties are called the *chemical elements*, and they help massively in the task of simplifying our understanding of the world around us. There was no guarantee, before we knew of these elements, that all matter could be dissected and categorized into such a relatively small number of fundamental constituents—compare this, for example, to the profusion of species we find in the living world, where there are more than 300,000 known varieties (and probably more than that still unknown) of beetles alone. So the manageably limited list of chemical elements is something to be grateful for.

All the same, it's a dauntingly long list for new students of chemistry. They might have already heard of elements like carbon and oxygen, but scandium? Praseodymium? Even pronouncing the names of some of these elements can be a challenge, let alone remembering anything about them or finding the motivation to do so.

That's where history might help. The elements were discovered only very gradually over time—from about 1730, at a surprisingly steady rate of about one every two or three years, with occasional bursts or hiatuses. This happened through no concerted program of seeking (at least not until the past several decades, when any newly discovered elements have had to be deliberately *human-made*). It was a haphazard business: scientists and technologists might find a previously unknown element in an obscure mineral, say, or by splitting sunlight into a spectrum to seek a telltale

gap where some new element has stripped a color from it, or by liquefying and distilling air to find tiny quantities of rare gases. These tales of discovery are like biographies, and they can make the elements seem less like a random collection of obscure nobodies and more like characters in the long and continuing saga of how we have tried to comprehend and manipulate our surroundings. To chemists, they genuinely come to acquire personalities: helpful or recalcitrant, intriguing or dull, friendly or hazardous. That chemists regularly conduct polls of their "favorite elements" can seem unutterably nerdy—until you get to know the elements yourself, in which case you will almost certainly find that you have developed preferences and aversions of your own.

Some elements have proved immensely useful: as crucial ingredients in drugs or other medical agents, for example, or for making new materials that are harder, stronger, shinier, better conductors of electricity, and so on. Some have brightly colored compounds (combinations with other elements) that are valuable as pigments and dyes. Some are sources of energy, or essential nutrients for health, or refrigerants that can keep things colder than deep space. Their properties and uses have even elevated a few elements to privileged inclusion in the cultural lexicon: opportunities are golden, clouds have silver linings, suggestions go down like lead balloons, opponents are crushed with an iron fist, provocateurs are denied the oxygen of publicity. We might speak of the sodium glare of streetlights, of hydrogen bombs, of nickel-and-dime stores, magnesium flares, with little if any understanding of why those particular elements are being invoked. The very notion of an "element" itself connotes a

OPPOSITE: A sage holding the tablet of ancient alchemical knowledge. From a later transcript of Muhammed ibn Umail al-Tamîmî's *Al-mâ' Al-waraqî* (The Silvery Water), Baghdad, ca. 1339, Topkapi Sarayi Ahmet III Library, Istanbul.

fundamental principle beyond chemistry, for we speak of the elements of law, of mathematics (the topic of the ancient Greek thinker Euclid's treatise *The Elements*), of language, of cookery.

All this implies that charting the history of the discovery of the chemical elements is more than an account of the development of chemistry as a science. It also offers us a view of how we have come to understand the natural world, including our own constitution. Furthermore, it shows how this knowledge has accompanied the evolution of our technologies and crafts—and "accompany" is the right word to use, because this narrative challenges the common but inaccurate view that science always moves from discovery to application. Often it is the reverse: practical concerns (such as mining or manufacturing) generate questions and challenges that lead to fresh discoveries. We can see too how scientific discovery is not some impersonal and inexorable process, but depends instead on the motivations, capabilities, and sometimes the idiosyncrasies of individual people: it requires determination, imagination, and ambition as well as insight and—never underestimate this—a substantial dash of good luck.

It's an unavoidable fact that histories of this kind must dwell, especially in the past several centuries, to a degree that we now rightly find uncomfortable, on the exploits and achievements of men of European heritage. Not only was it very difficult until recent times for women to gain entry into scientific institutions, but even those few who did often faced intense discrimination and prejudice. Marie Curie, for example, who did most of the work in finding the elements radium and polonium at the end of the nineteenth century, was nearly overlooked when

Periodische Gesetzmässigkeit der Elemente nach Mendelejeff.

Reihen	Gruppe I R^2O	Gruppe II RO	Gruppe III R^2O^3	Gruppe IV RH^4 RO^2	Gruppe V RH^3 R^2O^5	Gruppe VI RH^2 RO^3	Gruppe VII RH R^2O^7	Gruppe VIII RO^4
1	H=1							
2	Li=7	Be=9,08	B=11	C=12	N=14	O=16	F=19	
3	Na=23	Mg=24	Al=27,04	Si=28	P=31	S=32	Cl=35,37	
4	K=39	Ca=40	Sc=44	Ti=50,25	V=51,1	Cr=52,45	Mn=54,8	Fe=56, Co=58,6 Ni=58,6, Cu=63
5	(Cu=63)	Zn=65	Ga=68	Ge=72	As=75	Se=78,87	Br=79,76	
6	Rb=85	Sr=87,3	Yt=89,6	Zr=90	Nb=94	Mo=96	-=100	Ru=103,5, Rh=104 Pd=106, Ag=107,6
7	(Ag=107,6)	Cd=111,7	In=113,4	Sn=117,4	Sb=120	Te=126	J=126,5	
8	Cs=133	Ba=136,8	La=138,5	Ce=141,2	Di=145	-	-	- - - -
9	(-)	-	-					
10	-	-	Er=166	-	Ta=182	W=184	-	Os=191,12, Jr=192,6 Pt=194, Au=196
11	(Au=196)	Hg=200	Tl=204	Pb=206,4	Bi=207,5	-	-	
12	-	-		Th=232		U=240	-	- - - -

Verlag v. Lenoir & Forster, Chem-Physikal. Institut, Wien II. Waaggasse 5.

Lith. v. Scherrer & Northammer Wien, IX. Hofzst. 31.

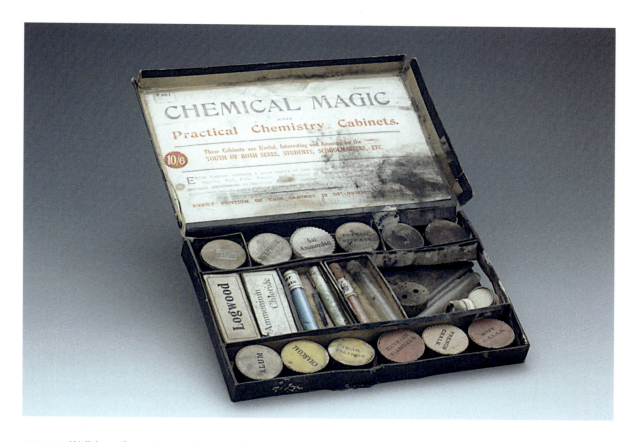

OPPOSITE: Wallchart of an early periodic table of the elements according to Mendeleev, 1893, Yoshida-South Library, Kyoto University.

ABOVE: *Chemical Magic and Practical Chemistry Cabinet* produced by F. Kingsley, London, ca. 1920, History of Science Museum, Oxford.

the work was rewarded by the 1903 Nobel Prize for Physics. Initially, the award was going to acknowledge only her husband and collaborator Pierre, before, forewarned, he objected. Similarly, Marie-Anne Paulze Lavoisier's contributions to the work of her husband, the eminent eighteenth-century French chemist Antoine Lavoisier, were long considered little more than wifely duties rather than those of a scientific collaborator. Even as late as the 1950s, the American nuclear chemist Darleane Hoffman, who made vital contributions to the discoveries of new, heavy radioactive elements, was told when she arrived at Los Alamos National Laboratory to lead a new team that there must

be some mistake because "We don't hire women in that division."

Why, meanwhile, people of color feature rather little in this story is a question fraught with the entire history of Western global dominance and exploitation since early modern times, as well as the prejudices and systematic biases that still lead to their under-representation in the sciences today. It isn't clear for how much longer the story of element discovery will or, for that matter, can continue—but the rise of scientific excellence in Asia, at least, might lead us to expect as well as to hope that, if it does, then it will feature significantly more cultural richness and diversity.

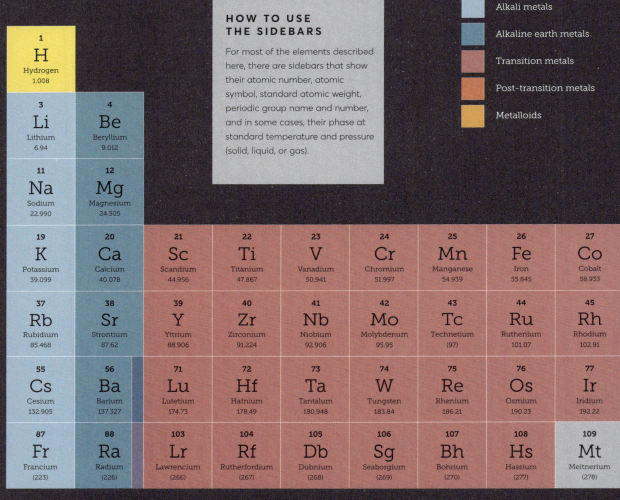

HOW TO USE THE SIDEBARS

For most of the elements described here, there are sidebars that show their atomic number, atomic symbol, standard atomic weight, periodic group name and number, and in some cases, their phase at standard temperature and pressure (solid, liquid, or gas).

- Alkali metals
- Alkaline earth metals
- Transition metals
- Post-transition metals
- Metalloids

1 H Hydrogen 1.008

3 Li Lithium 6.94
4 Be Beryllium 9.012

11 Na Sodium 22.990
12 Mg Magnesium 24.305

19 K Potassium 39.099
20 Ca Calcium 40.078
21 Sc Scandium 44.956
22 Ti Titanium 47.867
23 V Vanadium 50.941
24 Cr Chromium 51.997
25 Mn Manganese 54.939
26 Fe Iron 55.845
27 Co Cobalt 58.933

37 Rb Rubidium 85.468
38 Sr Strontium 87.62
39 Y Yttrium 88.906
40 Zr Zirconium 91.224
41 Nb Niobium 92.906
42 Mo Molybdenum 95.95
43 Tc Technetium (97)
44 Ru Ruthenium 101.07
45 Rh Rhodium 102.91

55 Cs Cesium 132.905
56 Ba Barium 137.327
71 Lu Lutetium 174.73
72 Hf Hafnium 178.49
73 Ta Tantalum 180.948
74 W Tungsten 183.84
75 Re Rhenium 186.21
76 Os Osmium 190.23
77 Ir Iridium 192.22

87 Fr Francium (223)
88 Ra Radium (226)
103 Lr Lawrencium (266)
104 Rf Rutherfordium (267)
105 Db Dubnium (268)
106 Sg Seaborgium (269)
107 Bh Bohrium (270)
108 Hs Hassium (277)
109 Mt Meitnerium (278)

Superheavy elements

57 La Lanthanum 138.91
58 Ce Cerium 140.12
59 Pr Praseodymium 140.91
60 Nd Neodymium 144.24
61 Pm Promethium (145)
62 Sm Samarium 150.36
63 Eu Europium 151.96

89 Ac Actinium (227)
90 Th Thorium 232.04
91 Pa Protactinium 238.03
92 U Uranium 238.03
93 Np Neptunium (237)
94 Pu Plutonium (244)
95 Am Americium (243)

Legend

- **Lanthanides**
- **Actinides**
- **Other non-metals**
- **Halogens**
- **Noble gases**
- **Not yet sufficiently characterized**

					2 **He** Helium 4.003
5 **B** Boron 10.81	6 **C** Carbon 12.011	7 **N** Nitrogen 14.007	8 **O** Oxygen 15.999	9 **F** Fluorine 18.998	10 **Ne** Neon 20.180
13 **Al** Aluminum 26.982	14 **Si** Silicon 28.085	15 **P** Phosphorus 30.974	16 **S** Sulfur 32.06	17 **Cl** Chlorine 35.45	18 **Ar** Argon 39.948

28 **Ni** Nickel 58.693	29 **Cu** Copper 63.546	30 **Zn** Zinc 65.38	31 **Ga** Gallium 69.723	32 **Ge** Germanium 72.630	33 **As** Arsenic 74.922	34 **Se** Selenium 78.971	35 **Br** Bromine 79.904	36 **Kr** Krypton 83.798
46 **Pd** Palladium 106.42	47 **Ag** Silver 107.87	48 **Cd** Cadmium 112.41	49 **In** Indium 114.82	50 **Sn** Tin 118.71	51 **Sb** Antimony 121.76	52 **Te** Tellurium 127.60	53 **I** Iodine 126.90	54 **Xe** Xenon 131.29
78 **Pt** Platinum 195.08	79 **Au** Gold 196.97	80 **Hg** Mercury 200.59	81 **Tl** Thallium 204.38	82 **Pb** Lead 207.2	83 **Bi** Bismuth 208.98	84 **Po** Polonium (209)	85 **At** Astatine (210)	86 **Rn** Radon (222)
110 **Ds** Darmstadtium (281)	111 **Rg** Roentgenium (282)	112 **Cn** Copernicium (285)	113 **Nh** Nihonium (286)	114 **Fl** Flerovium (289)	115 **Mc** Moscovium (290)	116 **Lv** Livermorium (293)	117 **Ts** Tennessine (294)	118 **Og** Oganesson (294)

64 **Gd** Gadolinium 157.25	65 **Tb** Terbium 158.93	66 **Dy** Dysprosium 162.50	67 **Ho** Holmium 164.93	68 **Er** Erbium 167.26	69 **Tm** Thulium 168.93	70 **Yb** Ytterbium 173.05
96 **Cm** Curium (247)	97 **Bk** Berkelium (247)	98 **Cf** Californium (251)	99 **Es** Einsteinium (252)	100 **Fm** Fermium (257)	101 **Md** Mendelevium (258)	102 **No** Nobelium (259)

CHAPTER ONE

THE CLASSICAL ELEMENTS

LEFT: The legendary sage or deity known as Hermes Trismegistus teaching the Egyptian astronomer Ptolemy the World System. Silver plate with relief decoration, AD 500–600, The J. Paul Getty Museum, Villa Collection, Malibu, California.

THE CLASSICAL ELEMENTS

"The body of the world," Plato wrote around 360 BC in his wide-ranging philosophical treatise *Timaeus*, "is composed of four elementary constituents, earth, air, fire and water, the whole available amount of which is used up in its composition." In other words, what we can see around us is all there is of this elemental stuff.

This quartet of elements is often portrayed as the universal scheme in the ancient world. But it wasn't really. The four-element system was formulated in the fifth century BC by Empedocles, a philosopher who is surrounded by exotic stories. Some say he was a magician who could raise the dead, while legend has it that, deciding he was an immortal god, he died by leaping into the volcanic maw of Mount Etna. As with so many accounts of people who lived before reliable historical records, such tales should be taken with a pinch of salt.

BELOW: The four elements. Detail from *Various Verse Treatises On Moral Subjects and Natural History*, Italian, 1481, Harley 3577 manuscript, The British Library, London.

Although Empedocles's elemental system persisted—partly thanks to its endorsement by the heavyweights Aristotle and Plato—in medieval Western tradition and beyond, there were several dissenting views, even among the Greek philosophers, about what the world is made of. That's hardly surprising because the answer is not obvious, nor is it easy to find out. But two principles seem to have guided these efforts. The first is that the fabrics of the world have rather diverse properties: some are solid, some fluid, some airy. Of course, we can make finer distinctions too: there are soft and sticky substances (like mud), say, as well as tough but pliable ones (like wood). They have different colors, tastes, smells. Yet many of the Greek philosophers drew only the most basic distinctions when trying to figure out what the fundamental elements were. Colors, for example, were superficial and could change—look at the way copper tarnishes to greenish verdigris. But solidity was shared by any substance that had a large degree of "earthiness."

The second guiding principle was that substances can be changed. Burn a log and much of it seems to vanish into the air, leaving an earthy residue of ash. Copper and iron can be melted into flowing form. So understanding the elements wasn't just a quest to describe a static, unchanging world; it also had to account for the transformations that we see around us.

It's tempting to imagine that ancient schemes of the elements arose from the same search for simplicity that made later chemists draw up a periodic table of elements and explain it with a unified view of what atoms are, or that led modern physicists to develop theories about the small family of fundamental particles that make up atoms. Maybe this impulse to find conceptual unity did indeed play a part—people have always found it helpful to try to break down complicated things and processes into simpler ones that are easier to grasp. That, after all, is a big part of the scientific enterprise. But the quest to understand the elements was also practically motivated. What was going on when bread was baked, when mortar set between the bricks, when a ceramic glaze developed a glossy hardness in a kiln? As we survey the history of the discovery of elements, never forget this: many of these discoveries have come about not because scientists, artisans, and technologists were looking for them, but because they were trying to make something useful. Chemistry is, and always has been, primarily an art of making—and if we want to know what the elements are, it's because it is always useful to appreciate your ingredients.

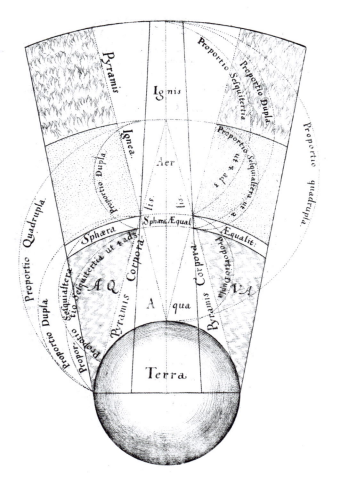

ABOVE: The four classical elements—*terra* (earth), *aqua* (water), *aer* (air), and *ignis* (fire)—depicted as arranged in concentric cosmic spheres. Robert Fludd's *Utriusque Cosmi Maioris Scilicet et Minoris Metaphysica, Physica Atque Technica Historia*, Oppenheim: J. T. de Bry, 1617, Getty Research Institute, Los Angeles.

PROTE HYLE

"Most of the early philosophers believed that the essence of all things could be reduced to material principles." This sounds like a quotation from a history book about the ancient view of the world. Yet it was actually written in the fourth century BC by Aristotle, for whom what the philosophers had said two hundred years earlier was already history. "That from which all things come to be, the original state of their generation…that is what they hold to be the element and the principle of things," he went on. "While the state of the substance can change, the substance itself remains." What

BELOW: Anaximander of Miletus holding a sundial. Early third century AD mosaic, Rheinisches Landesmuseum, Trier.

Aristotle is getting at here is the idea that ultimately there is only one primordial substance, from which everything else then arises.

If that were true, however, this would be a very short book indeed. Still, the picture that Aristotle paints is not so very different from the one most scientists believe in today. Our universe began with the Big Bang, when space and time and all it contains sprang from a seed so tiny and packed with energy that even our best physical theories can't describe it. What we do know is that there wasn't a lot of room for variety in that almost infinitesimally small bubble of space-time. All the distinctions we see today—between atoms of different elements, between the fundamental particles within atoms, and the forces through which they feel one another—must be erased if we rewind to those first instants of creation. That's about as much unity as we can imagine, and we don't have theories to tell us what it was like.

Yet Aristotle's primordial substance—which is often referred to in Greek as the *prote hyle* (first matter) or, later in Latin, the *prima materia*—wasn't anything quite as exotic and unimaginable as this. The idea was that the Creator—and the Greek philosophers did not doubt there must have been some cosmic creator, even if they didn't exactly imagine a God in the Judaeo-Christian sense— didn't work with four elements, but with just one, out of which others somehow arose. It's not easy today to grasp what this means. We tend to think of an element as *stuff*, as some substance we can see and hold. But the ancient Greeks often spoke of the prote hyle as a kind of "material cause": it brings into being all the substances we see around us. Not only can we not see this primal stuff, but it's not clear whether it is even *seeable*—any more than we can expect to look into the heart of the Big Bang and discern what is there.

The idea of prote hyle is often linked to one of the earliest traditions in Greek philosophy, known

as the Milesian (or Ionian) school, which flourished in the seventh century BC. The earliest known member of this school was Thales of Miletus, that location being a city on the west coast of Anatolia in modern-day Turkey. Thales is, in fact, essentially the first Greek philosopher about whom we know anything at all, and even that comes from later, secondhand sources like Aristotle. Thales, as we'll see, had his own views about prote hyle, but his pupil and successor Anaximander summed up what it is about this concept that is so elusive. He called it *apeiron*, the "unlimited"—though he might as well have labeled it "don't ask," since he imagined it to be invisible, infinite, eternal, and unchangeable. Elements such as earth, air, fire, and water come from apeiron by an "eternal movement," which brings about a separation of opposite qualities: hot separates from cold, say, and dry from moist. It's perhaps a little fanciful, but also irresistible, to see here too a presentiment of the way modern physicists believe the profusion of particles and forces we see around us stemmed from processes of separation that they call "symmetry breaking" in the very early universe, through which a thing that was initially uniform transforms spontaneously into two different varieties.

In any case, the idea that the classical elements are both united and differentiated by their properties was popular with many thinkers after Anaximander. It was precisely because (unlike apeiron) elements such as water and earth have properties at all that we can experience them: we feel the wetness and coldness of water, for example. For Aristotle, one element can be transformed to another by a switch in their properties: if wet, cold water loses its wetness, it becomes dry, cold earth—which in turn becomes fire when the coldness is changed to hotness. An elemental system such as this might sound crude today and in some sense "wrong"—but it was a start toward making sense of how the physical world works.

If you think that Anaximander's prote hyle sounds rather vague and elusive, just consider what the philosopher Pythagoras made of it all. He lived on the Aegean island of Samos in the fifth century BC, and he and his followers took the view that what

ABOVE: The origin of matter: "In one body—cold fighting heat, wet, dry." From Michel de Marolles's *Tables of the Temple of the Muses*, Amsterdam: A. Wolfgank, 1676, Book 1, University of Illinois Urbana-Champaign.

was truly fundamental in the world was not a substance—whether it was tangible and visible or not—but numbers. They regarded numbers as concrete and real, almost as if they are elemental "shapes"—one is a point, two a line, three a plane—from which all objects are built. And here, too, there are anticipations of the modern view in which all matter and forces are described by physicists in purely mathematical terms: here the stuff that chemists handle and transform seems to have vanished into abstraction.

WATER

Thales of Miletus decided that the primal substance from which all others come—the unique generative element—was water. Perhaps that sounds unlikely, but you can understand his reasoning. In the ancient world, water was the only known substance that could adopt all the states of matter: although most familiar as the liquid that fills the ocean basins and rushes through the channels of streams and rivers, it could also freeze solid as ice and evaporate into "air" (today we would say into vapor or gas). It also seems to be the essential source of all life: Thales had witnessed how vital the seasonal flooding of the Nile river was to the replenishment of the fertile alluvial deposits of its delta. Aristotle believed Thales was also influenced by the view that all foods are moist and that seeds germinate from moisture.

In Thales's view, the other classical elements—air, fire, and earth—were all derived from water. The first two are airy "exhalations" of it, while earth appears out of water as a kind of sediment—not only are such tiny particles almost always present in river water, but a solid crust of salt remains when seawater evaporates. The Greek-Roman physician Galen claimed that one of Thales's original texts—now lost, if Galen was telling the truth—stated that the four elements "mix into one another by combination, solidification, and incorporation of things of the world."

Doesn't the evidence for leaping to such a far-reaching conclusion seem just a little…thin? Well, yes, by today's standards, when scientists propose ideas based on a careful observation of data and then test them experimentally to see if they stand up. But there was no such conception of science in the ancient world, and indeed it didn't really start to cohere until the seventeenth century. Yet by suggesting that water was the fundamental element, the basic component of all things, Thales was saying something important for the development of thought. First, this idea doesn't ascribe anything to the arbitrary whims of the gods or to causes that we might now regard as superstitious; it's a thoroughly rational idea. (Thales apparently thought that some divine agency provides the guiding intellect that fashions things out of water, but we can hardly hold that against him; many people today view God's role in a similar manner.) And it's also a simplifying idea, an attempt to explain many observed facts with a single underlying one. It's not yet science, but it is the kind of thinking science was going to need.

As we saw from the example of Anaximander, Thales's belief that water is the primordial element wasn't shared even by some of his followers. Others, however, embraced the idea. Hippon of Samos, who was not just a contemporary but virtually a neighbor of Pythagoras, thought that both water *and* fire were the most basic elements. And, in fact, the primacy of water was even being asserted as late as the

LEFT: A Greek *clepsydra* (water clock), made from clay, late fifth century BC. This one held two choes (6.4 quarts) and took six minutes to empty. Museum of the Ancient Agora, Athens.

seventeenth century, when the Flemish physician Jan Baptista van Helmont claimed to have *demonstrated* that everything was made of water.

Van Helmont was carrying out an experiment that had been proposed two hundred years earlier by the German cardinal and natural philosopher Nicholas of Cusa. Imagine, Nicholas had said, that you took a pot filled with earth, planted some herb seeds in it, and watered them daily. Before long you'd have some fine plants. But from where had they got their substance? Not from the earth, of which there would be as much as you started with. No, all you'd added was water—so surely *that* must be what the herbs were made from.

It's not a difficult experiment, and in a way many people did it over the ages to season their stews. But van Helmont conducted it as a scientist: by weighing the soil at the outset and at the end, along with the plant (a willow sapling) that he grew for five years. He even covered the pot with a metal lid, pierced with holes to let the air in but to keep dust out. And, in the end, he said, "one hundred and sixty-four pounds of Wood, Barks, and Roots arose out of water onely." It didn't, but that could hardly have been obvious to van Helmont. Plants build their fabric from carbon dioxide gas captured from the air in a process powered by the energy of sunlight. Moisture is essential, but not the only ingredient. The clever chemistry involved in *photosynthesis* (literally, "making with light") was only grasped in the twentieth century, and so we'd be well advised to be humble in giving due appreciation to the efforts of ancient philosophers to understand what are the elements of all things.

LEFT: The ancients watering with a "thumb pot." Title page from Charles Estienne's *Maison Rustique*, London: Adam Islip for John Bill, 1616, Wellcome Collection, London.

AIR

Just as Thales was succeeded by Anaximander as the leading light of the Milesian school of philosophy, so Anaximander had a protégé, named Anaximenes. He too had other ideas about the nature of prote hyle, the primordial element. It's not water, he said, and neither was he content to fall back on Anaximander's vague notion of apeiron. No—it is air. That might in itself seem a rather arbitrary substitution, but Anaximenes saw logic to it. The creation of the world was regarded back then as a process by which structure and substance emerged from a primordial Chaos—and what is less ordered and more chaotic than swirling, restless air? (The very word "gas," for which air is the familiar archetype just as water is the archetypal liquid, is derived from "chaos." The word is said to have been coined in the seventeenth century by Jan

Baptista van Helmont.) Anaximenes imagined how that process of emergence happened: through what we today call condensation, in which gas (here, air) collapses to a dense substance. First, he said, air becomes water, and then as the density increases, earth and stone. This happens through a loss of heat—or as philosophers might have said then, by the action of cold. Conversely, air can undergo rarefaction (becoming even more tenuous) by raising its temperature, whereupon it becomes fire. In Anaximenes's "air-first" cosmology there's a rational, even mechanical, description of how all things came into being.

That's not to say there wasn't a mystical side to this belief, too: Anaximenes was said to have thought that ultimately air was the stuff of the supreme being. And given that air was at the time still thought to be without any mass or "body" itself, his idea wasn't so different to Anaximander's notion of elusive apeiron anyway.

Empedocles is usually attributed, in the fifth century BC, with realizing that air really does consist of "stuff"—you might even call it the discovery of air in the modern sense. It's said that he demonstrated this in an experiment using a water clock—what the Greeks called a *clepsydra*. There are several different types of water clock, but all work by measuring time according to how long it takes water to flow into or out of a vessel with holes or openings in it. One form is an inverted cone that drains through a small hole in its downward-pointing apex; in another, the passage of time is determined by how long it takes such a cone to fill and sink when placed in water. Empedocles's experiment seems to have involved blocking the outlet of a *clepsydra* with a finger so that water can't fill it completely when it is immersed, due to the air bubble trapped inside. When the finger is removed, the air bubbles out and the vessel sinks completely. So the air can't just be "nothing," since it has to get out of the vessel before water can enter.

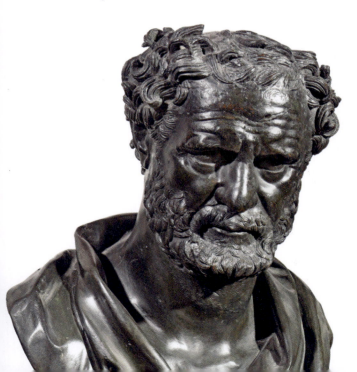

BELOW: Bronze bust of Empedocles, second half of third century BC, Villa of the Papyri, Herculaneum, National Archeological Museum of Naples.

It's sometimes said that this was the first scientific experiment ever recorded. That's not a very meaningful claim, however. For one thing, a true experiment is a test to see if an idea is right or wrong (or perhaps just to gather information about an unexplained phenomenon). But it's doubtful that Empedocles would have changed his view if things hadn't turned out the way he expected; like most "experiments" in ancient times, this was more of a demonstration. And, in any case, it's very likely that it never happened at all; Empedocles simply described a girl carrying out the operation, and so he was probably just explaining what *would* happen—which, with its bubbles emerging from a submerged container, was probably pretty familiar to his audience. All the same, it was

widely accepted from his time onward that air was indeed a physical substance, albeit one you couldn't feel, see, or taste.

Not, that is, unless the air moves. "The air round the earth is necessarily all of it in motion," wrote Aristotle, and this was the origin of the winds. As particles of air grow heavy, he said, they lose their warmth and sink—while fire may mix with air and cause it to rise. This interplay of air and fire, cooling and warming, produces the roiling of the atmosphere: an impressive presentiment of the modern view that convection currents and the differences in temperature, pressure, and humidity of the air may elicit everything from a gentle, flag-ruffling breeze to a raging hurricane.

RIGHT: Empedocles's four elements. Colored woodcut from Lucretius's *De Rerum Natura*, first century BC, printed in Brescia by Thomas Ferrandus, 1473–1474, John Rylands Library, University of Manchester.

FIRE

Three of the classical elements of Empedocles are representatives of the three states of matter: earth is solid, water liquid, and air gas. But where, then, does fire sit? It's the odd one out, for sure. Today, we know that fire is not a substance, but a process: it is what results when combustible substances burn. The bright, flickering flames of a gas or wood fire are composed of tiny soot particles so hot that they glow like the filament of a light bulb. They condense from a gaseous mixture of many different chemical substances, mostly carbon-based molecules broken up into small fragments or even into solitary atoms. The edge of a flame marks the place where the temperature falls too low for the soot particles to emit light. So the truth is that fire—which is to say, a flame—is extremely complicated, and even now the chemistry involved is not completely understood.

It's not hard to see why ancient philosophers thought there was something special and unique about fire. For there really is. It seems not just to hold but to generate heat—and it is also a source of light. Both of those things have been immensely valuable to humankind since long before recorded history. Some anthropologists argue that it was not so much the discovery of fire (at least 400,000 years ago) but of cooking that marked a turning point in human prehistory: the calorific boost of easily digested cooked meat fueled the development of bigger brains and freed up time that was otherwise devoted to chewing and digesting. With fire, our ancestors could also fend off the ice-age chill, keep predatory beasts at bay, and stay active and socializing after night fell.

One view is that, by making fire an element along with the other three, Empedocles's scheme embraced not just states of matter, but those other two vital aspects of the physical world: heat and light. Neither of these was close to being understood until the late nineteenth century, but elemental fire offered at least some reassurance that our intellect and worldview could contain them.

Given the importance of fire, it's perhaps not surprising that it too has been proposed as the primordial substance. That was the view of Heraclitus of Ephesus, a city in modern-day Turkey, around 500 BC. In one sense, he was simply choosing to focus on a different stage in the progression also identified by Thales and Anaximenes by which elements are transformed one into another in processes of condensation and rarefaction: fire may condense to water and then further to earth. For Heraclitus, however, those processes reflected his view that the *cosmos*—a term that first appears in his writings—was constantly changing, always in flux. It was Heraclitus who expressed the idea that one can never step in the same river twice. Without change, nothing could exist, and Heraclitus saw this as a consequence of the play of opposing forces: "all things happen according to strife and necessity," and only out of strife and conflict does harmony emerge. Something is always burning somewhere.

This was a fitting position given that fire was, at that time and long after, the ancient chemist's main and almost sole agent of transformation. It was the only means of inducing a change from one thing to another: smelting metals, baking bread, melting sand and soda into glass. The practical arts of chemistry were born of fire.

OPPOSITE: Claudio de Domenico Celentano di Valle Nove's *Book of Alchemical Formulas*, Naples, 1606, Getty Research Institute, Los Angeles.

Hec est Virgo Pascalis que primam vir-
tutem tenet in capitis suis et est
herba multum vigens in puteis

quatuor sunt spiritus, due
facies sed ista sunt qua-
tuor elementa, nam
distillationem habes aqua
et aerem calcinatione
habes ignem et terram
et terra suam frigiditatem
aque prestat et aqua
suam humiditatem
aeri donat, aer suam
cahiditatem igni commu-
nicat

Sic circulantur
vicem element
quatuor sunt s
due facies, in ist
et sic ignis vir
aere, aer de n
to aqua, aqua
trimeno terre,
Lapis ex omni
mentis puris
vivit

Estas

Autumnus

Tota scientia   Lapidis manifesta

verte oculos   ad igne
ibi ista   tus

Aperi oculos   ad igne
ibi   tempo

Lapis

Ego sum esaltata super
quarum una est in
debet poni in lapide

circulos mundi, ubi quatuor facies habentes unus par
hus alius in aere alia in cavernis, alia in saxis vel con
sam solem

SOLIDS: EARTH, WOOD, METAL

If you are now thinking that surely one of these ancient philosophers must have made the fourth element of the classical pantheon—earth—their prote hyle, you'd be right. Xenophanes of Colophon, who lived during the late sixth and early fifth centuries BC and founded the so-called Eleatic school of philosophers, has been attributed with saying "Everything is born of earth and everything returns to earth"—in which we can hear a premonition of the Christian ritual phrase: "Ashes to ashes, dust to dust." And, after all, doesn't earth seem the most likely candidate for a primordial matter, given that, like most of the objects around us, it is solid, visible, and tangible? We have even named our own world after this substance.

Yet even ancient sources are divided as to whether Xenophanes really took the view that earth was the basis of all matter. Some, such as Galen, say

LEFT: Second day of Creation (Genesis 1:1-8), God divides the water from the earth. From *Bible Pictures* by William de Brailes, Oxford, French manuscript on parchment, ca. 1250. The Walters Art Museum, Baltimore.

OPPOSITE: The Universal Chart of the Eight Trigrams showing the *Wu Xing* (Five Elements) at the center. From Wu Weizhen's *Wanshou Xianshu* (The Immortals' Book of Longevity), Ming Dynasty, Wellcome Collection, London.

he asserted that there were *two* basic elements: earth and water. He was certainly interested in both of these; he discussed the water cycle and the formation of clouds from moisture drawn up from the sea by the heat of the Sun, long before Aristotle wrote about such natural processes in his great text on weather and the Earth, *Meteorologica*. Furthermore, Xenophanes's notion that the world emerged from the interplay of earth and water mirrors the Christian story of Genesis: "And God called the dry land Earth; and the gathering together of the waters called he Seas."

All the same, the world view of Xenophanes and the Eleatic school contrasted with Heraclitus's

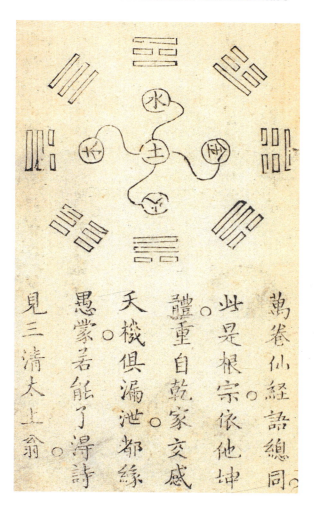

cosmos of flux, stressing instead concepts of permanence and unity. That's what you might expect from someone inclined to set solidity at the heart of the elements.

Yet earth wasn't the only common solid substance in the ancient world. In China, philosophers believed there were five fundamental substances: water, fire, earth, wood, and metal (the *Wu Xing*). These corresponded to the *five* cardinal directions in Chinese thought: not just north, east, south, and west, but also the center. In this scheme, earth occupies the central position and represents the coming together of all the elements. A Chinese Han-Dynasty treatise of around 135 BC says that "Earth has its place in the center and is the rich soil of Heaven…Earth is what brings these Five Elements and Four Seasons all together…if they did not rely on Earth in the center, they would all collapse."

The Chinese five-element system was first laid out clearly in the third century BC by the philosopher Zou Yan—who, Confucius and Lao Tzu notwithstanding, has been described as the real founder of all Chinese scientific thought. Just as the seasons change one to another, so the five elements can be transformed in a cyclic view of the universe that reflects a faith in the process of death and rebirth. This succession of substances is a central concept in alchemy, underpinning the idea that metals can be transmuted so that lead might be turned to gold. For Chinese alchemists in particular, this link between transmutation and the cycle of life connected their chemical manipulations to the possibility of sustaining human life by preparing elixirs. All this transformation operates according to the balance of the opposed cosmic forces of *yin* and *yang*, which play a similar role to the Love and Strife, the mixing and separation, invoked by Empedocles and other Greeks. Again, without making too much of the parallel, it's impossible not to be struck by the echoes here of modern ideas of how the physical world gets its forms from basic substances and particles interacting via forces—and, in particular, by how the atoms from which all chemical elements are made are constituted from subatomic particles united in a delicate balance of electrical attraction and repulsion.

IN SEARCH OF THE ATOM

The word atom comes from the Greek *atomos*, which means "unsplittable." We know now that atoms *can* be split (as well as merged), and later we'll see how that process has given us many new elements. But even if atoms aren't the most fundamental units of matter, the concept of a chemical element only makes sense up to this degree of subdivision: pull matter apart beyond the level of the atom, and there can be no elements left.

It's both extraordinary and odd that the ancient Greeks—at least, some of them—decided that all substances must be made of atoms: that ultimately they have a graininess beyond which they can't be divided any more. This, after all, isn't our everyday experience. You can cut a piece of cheese smaller and smaller, and if there comes a limit to that process, it's only because your knife or your vision is too blunt. With a razor blade and magnifying glass you can do better, and with a microscope, better still. Why would anyone think there was a limit?

Yet Leucippus of Miletus, in the fifth century BC, did come to that conclusion. At least, so we're told—what little we do know about him comes from the accounts of others, and it's not even completely agreed where he was born. We know more about the philosopher said to be his pupil, Democritus, who is believed to have come up with the word *atomos* to describe these indivisible grains.

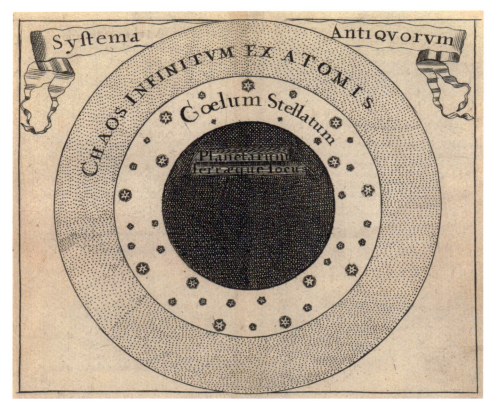

LEFT: The *Systema Antiquorum* or Democritean Universe. From John Seller's *Atlas Cælestis*, London: J. Seller, ca. 1675, Chart 23, Robert Gordon Map Collection, Stanford University Libraries, California.

OPPOSITE: The Platonic solids with their respective elements. Detail from Johannes Kepler's *Harmonices Mundi*, Linz: Johann Planck, Book V, 1619, Smithsonian Libraries, Washington DC.

This early atomistic theory may have been an attempt to reconcile the view of the Eleatic school that permanence sits at the heart of matter with the evident fact that—as anyone could see—change occurs. Perhaps change is nothing more than the rearrangement of imperishable, eternal atoms? Maybe too these various arrangements of just a few types of atom could explain how it is that the world contains a mere handful of elements yet innumerable varieties of substance? Aristotle compared this with the way in which just a small number of letters is needed to make an almost limitless number of words—an analogy that is spookily close to the metaphor chemists deploy today for how atoms combine to make a panoply of molecules and materials.

If, however, all stuff is composed of atoms, what sits between them? For Leucippus and Democritus, this was simply empty space: a void. Other philosophers thought it ridiculous to suppose that nothingness could exist; some felt that atoms must fill up all of space completely, while others argued that matter must be infinitely divisible after all, so that tiny grains could fill up any nooks between larger grains *ad infinitum*. Aristotle argued that if there were spaces between atoms, they'd be filled with air—which was all very well unless you accepted that air was an element like the others and therefore made of atoms, too.

What were atoms like? Democritus didn't say,

but Plato in the third century BC had his own ideas. Because he was convinced that the cosmos had been built by the Creator using principles of mathematical harmony and perfection, he decided that these atoms would have the shapes of the symmetrical, three-dimensional objects (polyhedra) that can be made from regular polygons: flat shapes in which all the sides and angles are equal. There are an infinite number of regular polygons, but only three of them can be the building blocks of polyhedra containing just one type of face: equilateral triangles, squares, and pentagons. And there are just five of the resulting polyhedra, now called the Platonic solids.

Plato said that four of these objects represent the shapes of the atoms of the four elements—and that those shapes help to explain the element's properties. Solid, stable earth is made by packing together cube-shaped particles, while the polyhedra with the fewest faces—the tetrahedron—is the most mobile and therefore the unit of fire. What's more, the tetrahedron has the sharpest points, which is why fire is so "penetrating." Air and water—respectively, an octahedron and icosahedron, both also made up from equilateral triangles—are intermediate states between this solidity and mobility.

"We must, of course," Plato wrote, "think of the individual units of all four bodies as being far too small to be visible, and only becoming visible when massed together in large numbers." Again, what's most impressive about these early ideas of the elements—wrong though they are—is that they attempt to explain why the stuff of the world behaves the way it does based on a theory of what they are made of at scales we can't see or (these philosophers believed) ever hope to see.

This makes it sound as though Plato shared Democritus's views about the atomic nature of matter, but that he made them geometric atoms. Yet that's not quite right. It's not easy to figure out how "real" Plato thought these atoms are, and he never even deigned to mention Democritus. But for Plato, all of reality as we know it had an ambiguous quality: it was, he suspected, just the shadow of something eternal, harmonious, and geometrical.

AETHER

What about that fifth Platonic solid? It is the dodecahedron, and its twelve faces are pentagons. Was there any place for it in the Platonic cosmos? There was, but not on Earth. "The gods used [it]," Plato wrote, "for embroidering the constellations on the whole heaven." It comes closest out of all the Platonic solids to resembling a sphere, the most perfect and symmetrical of shapes, and so it is the most suited to being the stuff of the eternal, perfect heavens. Aristotle adopted this idea and gave it a name: it was the "fifth element" or *quintessence*, which he also called the aether.

For Aristotle, the four classical elements have innate predispositions to move in certain directions: fire and air go up, water and earth go down (think of rain and the falling of a thrown stone). Aether did neither. Being perfect and outside the earthly realm, it reflected the behavior of the heavenly bodies

(which were made from it) by moving in circles. At a stroke, then, he explained why those objects—the Sun, Moon, planets, and stars—appear to rotate around the Earth. They do so because that's in the nature of their substance. This wasn't, to tell the truth, much of an explanation at all: the argument, like the motion, is circular.

This "fifth element" then was a rather makeshift idea. No one had ever seen it, and no one ever could: it wasn't possible to transmute any of the four earthly elements into aether. And the aether was intangible and invisible—in the figure of speech that stemmed from it, it was ethereal.

All the same, the idea was remarkably tenacious. It seemed to imply a fundamental divide between the Earth and the heavens—aether was governed by quite different rules from earthly matter. That idea persisted, more or less, right up until the early

LEFT: Ptolemaic model of the universe, with the Earth at center, surrounded by the other three elements. From Andreas Cellarius's *Harmonia Macrocosmica*, Amsterdam: Johannes Janssonius, 1660, Barry Lawrence Ruderman Map Collection, Stanford University Libraries, California.

seventeenth century, when observations using the newly invented telescope by Galileo and others showed that the Moon wasn't a perfectly smooth sphere, as Aristotle had insisted, but a rugged world like ours, with mountains and valleys. Very quickly, natural philosophers started to regard the heavens not as some remote, perfect, and inaccessible realm, but as just another part of the cosmos, not unlike a distant continent across the seas to where one day we might voyage. That picture came to seem inevitable once people began to accept the idea, proposed in the sixteenth century by Nicolaus Copernicus and supported by Galileo, that the Earth is not after all at the center of the cosmos, but is a planet like the others, circulating around the Sun.

Meanwhile, the idea of "ether" as a kind of very tenuous, gas-like substance led to it becoming a term used in chemistry to refer to a class of volatile (and pungent) liquids, which we now know to be made of carbon-based molecules. The most common ether, made from alcohol, was used in the nineteenth century as an anesthetic. That was quite a demotion for Aristotle's fifth essence.

At the same time, however, scientists clung to the idea that another kind of aether pervades the entire universe. Isaac Newton suggested in the early eighteenth century that a substance like this might carry the force of gravity acting between bodies. And in the nineteenth century, physicists thought that such a fluid—which couldn't be seen or directly detected—carried light waves, just as sound waves are vibrations of the air. They called it the *luminiferous* (light-bearing) aether. It was almost an unquestioned belief until attempts to detect this aether in the 1880s—by looking for the expected differences in the speed of light parallel and perpendicular to the motion of the Earth as it swept through the putative sea of aether—failed to show any sign of it. At first, some physicists tried to explain how the *luminiferous* aether could exist while being undetectable. But in 1905 Albert Einstein showed that the same mathematics could be used to describe how light travels through space without needing to invoke this light-bearing aether at all. The legacy of Aristotle's fifth element had finally come to an end.

RIGHT: Copernicus's heliocentric universe. From Andreas Cellarius's *Atlas Coelestis*, Amsterdam: Johannes Janssonius, 1660, Glen McLaughlin Map Collection, Stanford University Libraries, California.

CHAPTER TWO

THE ANTIQUE METALS

LEFT: Metalworking in ancient Egypt. From the tomb of the vizier Rekhmira, Eighteenth Dynasty (1549–1292 BC), Abd el-Qurna, Theban Necropolis, Luxor.

THE ANTIQUE METALS

The traditional names for different eras in early human history—the Stone Age, Bronze Age, Iron Age—are reminders of the transformative potential of materials. Having new substances from which to make tools and other implements may totally alter what we can achieve with them—perhaps with consequences for how we organize our societies and how we think about our relationship with the world. The latest coining for our present times as the Silicon Age invokes another element and makes it clearer than ever how material culture infuses our lives and potentially creates nothing less than a new reality.

It's notable, then, that two ancient eras—the Bronze Age and the Iron Age— are named after metals—specifically, for metals that, to be available in quantities sufficient to change culture, needed to be obtained by means of chemical technology. Bronze and iron were smelted from ores, and they exemplify the profound realization that the materials that may exist in the world are not limited to those that nature provides. One of the most important concepts in the history of civilization is often overlooked when the story of science is told: the notion of transformation, of taking a piece of the world and altering it, not just in shape but in its chemical nature. Sure, it was of immense significance that early humans figured out how to chip away at flint to produce tools—for hunting, for fighting with rivals, for carving wood and bone into useful and even artistic forms. But the production of metals was a change of a different order, prompting humans to wonder what more might be possible from the recombination of elements, which was then typically achieved by the agency of fire.

Needless to say, artisans of the Bronze and Iron Ages would not have understood what they were doing in these terms: that is, rearranging the chemical elements as we now understand them into new configurations. All they had to go on in developing ideas about how matter was constituted was readily observable properties: weight, color, hardness, and so on. It was small wonder, then, that many early thinkers supposed metals to be varieties of the same basic substance—that there was, to put it one way, a kind of Ur-metal of which gold, silver, iron, and so on were different manifestations, any one of which might perhaps be transformed into another. It would be unfair even to call such a view a mistake, given that it was consistent with the evidence then available. Such consistency between theory and what we can see is the best we can ever hope for from science, even today.

Ancient metallurgy was not a theoretical discipline anyway; it was a practical craft, and one that deserves our enormous respect. Metalworkers figured out through trial and error how to produce spectacular results: how to temper steel so that it held its edge, how to cast bronze and to adjust its qualities (such as brittleness and hue) by altering the mixture that went into its making. The quality of gold-working in the artifacts of ancient Egypt still commands our admiration. It was often the craftspeople, not the idea-smiths, who drove these technologies forward. In the first century AD, the Roman writer Pliny the Elder disapproved of

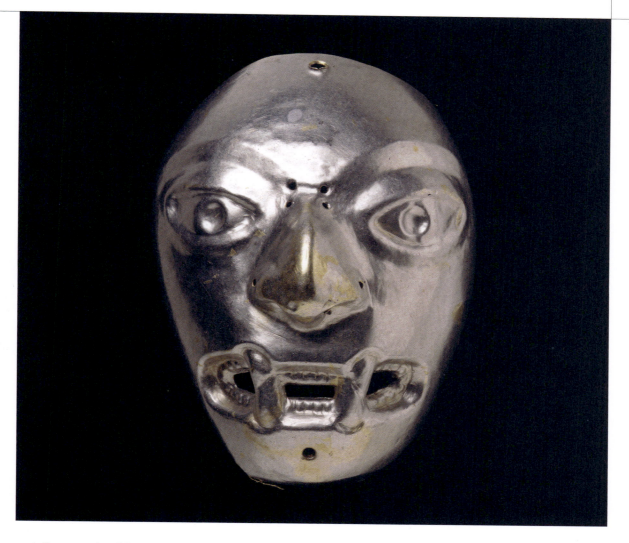

metallurgy, saying: "How innocent, how happy, nay more, how even luxurious would life be, if our desires did not go deeper than the surface of the earth, and were satisfied, in a word, with what is within our reach!" He wished that gold and silver could be "driven utterly from our lives." The Greek legend of King Midas warns where an excessive craving for gold is liable to lead.

Seeing now how mining and manufacturing have despoiled the environment, and how a lust for gold and silver has led to exploitation and enslavement of whole cultures, we can't perhaps help feeling some sympathy for Pliny's view. However, it seems to be part of human nature not to be satisfied with "what is within our reach," and increasing mastery

in transforming combinations of elements, which began in the era of the antique metals, has given us plenty that makes life not just more luxurious but—far more importantly—safer from disease and natural hazard. An ability to control the elements is a mixed blessing, to be sure; its pros and cons reflect the struggles within our own nature, where craving and desire battle with wisdom and restraint. In that respect, sadly, we seem to have advanced very little since antiquity.

COPPER, SILVER, AND GOLD

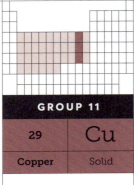

GROUP 11	
29	Cu
Copper	Solid

TRANSITION METAL

ATOMIC WEIGHT: 63.546

GROUP 11	
47	Ag
Silver	Solid

TRANSITION METAL

ATOMIC WEIGHT: 107.87

GROUP 11	
79	Au
Gold	Solid

TRANSITION METAL

ATOMIC WEIGHT: 196.97

You'll have probably heard of the Stone Age and Bronze Age, but what about the Chalcolithic or Copper Age? This period is considered to have bridged the most recent period of the Stone Age—the Neolithic—with the advent of bronze, an alloy of copper and tin. It stretches from around 4500 to 2000 BC, during which time there is widespread evidence of copper metallurgy in the Middle and Near East and in Europe.

The use of copper goes back even further. The earliest known artifacts made from this red metal are beads from northern Iraq, which have been dated to around 8700 BC; a copper bead from the region of modern-day Turkey has been dated to 500 years later. The people of this time didn't know how to extract copper from its ores, nor even how to melt and work the metal. For it can be found naturally in its "native" form—as the metal itself—and is soft enough to be beaten and worked without heating. The metal can crystallize from hot underground fluids rich in dissolved copper salts—called hydrothermal fluids—in places containing rich deposits of copper ore. Some of the most abundant deposits of native copper are found on the Keweenaw Peninsula, projecting into Lake Superior in present-day Michigan, where it was mined by Native Americans for thousands of years.

The conventional date for the start of the Bronze Age—3000–2500 BC—is a little misleading, as there is good evidence to suggest that bronze was being made considerably earlier. Tools cast from molten copper have been found in the central Balkans from the fifth millennium BC, and it seems that some of these didn't rely on native copper metal, but were smelted instead from copper ores such as malachite (copper carbonate) and chalcopyrite (copper-iron sulfide). The cultures of this region, in particular the Vinča in present-day Serbia, figured out that copper can be made harder by mixing it with tin—some of the earliest bronze objects come from

RIGHT: Copper figure of King Shulgi of Ur carrying a basket. From Nippur, Mesopotamia, ca. 2094–2047 BC, The Metropolitan Museum of Art, New York, Rogers Fund, 1959.

ABOVE: Cupids working as goldsmiths, *Triclinium* of The House of the Vettii, Pompeii, Italy, first century AD.

this time and place. The ancient Balkan metallurgists even seem to have controlled the proportions of the two metals, as well as the natural arsenic impurities, to give their artifacts a desirable golden hue. The smelting of copper and the creation of bronze was also practiced by the people of Mesopotamia and the cultures of the Indus Valley around the same period, although it's not clear who discovered this technology first or how the knowledge spread. The metal was mined on the island of Cyprus as the major source for the Greek and Roman cultures. The Romans, in fact, named the metal after the island, calling it *cuprum*, which later became the old English *coper*.

Copper was the first useful metal because it could be used to make bronze. Relatively hard and strong, bronze was used for making everyday items such as knives, tools, razors, and cutlery. Some of the common tools of later ages, such as the chisel, rasp, and sledgehammer, first appear in bronze. It was also used for ornamental and artistic objects, from jewelry to monumental statues, most famously for the 105-foot Colossus of Rhodes, depicting the

sun-god Helios, which was built around 292–280 BC to commemorate (ironically) the victory of the city of Rhodes over Cyprus itself. Of course, bronze was also used for weapons and armor. The alleged destruction of Troy, in modern-day Turkey, described by Homer in the *Iliad*, is commonly regarded as marking the end of the Bronze Age itself, although how much true history there is in Homer's epic account of the Trojan Wars remains a matter of debate even now.

A lot of the copper mined and smelted in ancient times was used for coinage—where it typically supplied the lowest currency denominations. For it has, of course, always been relegated to third place among the classic coinage metals, beneath silver and

BELOW: Silver, gold, and electrum (an alloy of gold and silver) coinage. From left: Silver tetradrachm, from Bostra, near Athens, 475–465 BC; Achaemenid gold daric, from the Persian Empire, 500–400 BC; Greek electrum hecte, from Cyzicus, Mysia, Asia Minor, 550–500 BC, The J. Paul Getty Museum, Los Angeles.

ABOVE: Jason and the Argonauts arriving in Colchis. Woodcut from Georgius Agricola's *De Re Metallica*, Basel: 1557, Book VIII, University of California Libraries.

gold. We take it for granted that these three metals are well suited to being used as agents of monetary value because they—silver and gold especially—retain their pleasing shine and don't tarnish easily. But this resistance to corrosion is rather unusual among metals, and it is why the coinage metals are the original members of a family known by the medieval name of the noble metals. Today that term has lost its association with royalty, and for chemists it connotes instead a lack of chemical reactivity. For copper, silver, and gold this property stems from the same source: all three occupy the same group in the periodic table, and their atoms share an arrangement of electrons that makes

them especially stable and slow to react with other compounds—such as those in air and moisture. The long-standing attribution of value to silver and gold, then, has a chemical explanation.

This unreactive character explains too why both silver and gold may also be found in nature in their native form. For gold in particular, this is the main source: it doesn't need smelting from an ore, but can simply be picked from the ground as nuggets, dug from bright veins, or sifted as sparkling alluvial grains from streams. Again, we don't know when this first happened, but it is a very ancient practice. There is evidence of the mining of gold in Armenia and Anatolia from before 5000 BC. Natural gold is not pure, but is typically an alloy with small amounts of silver. When the silver content is more than 20 percent, the metal looks more silvery than golden; the Greeks called this "white gold" or *electron*, which became *electrum* in Latin. It was harder than purer gold, and so made a more durable coinage metal. Much of the alluvial gold that could be sifted from the Pactolus river in ancient Lydia was, in fact, electrum—this was the river said in legend to have received its precious load after King Midas bathed in it to free himself from the wish for the "golden touch" he unwisely and greedily requested as a reward from the god Dionysius. Lydia was also the kingdom of the legendarily wealthy King Croesus, who ruled from around 561 to 547 BC; he replaced the electrum coins minted in Lydia for about a century with ones of pure gold and silver.

The lure of gold and silver

Alluvial gold was common in the natural waters of Asia Minor. According to the Roman writer Strabo, the people of Colchis, a kingdom between the Caucasus, Armenia, and the Black Sea, would collect gold on animal skins and the fleeces of sheep placed in the pools of springs: the origin, he said, of the legend of the Golden Fleece. This waterborne gold is released from veins or "lode" deposits when streams and rivers wash away the surrounding rock. Gold can be mined in greater abundance from the veins themselves, and gold mining was an important activity in ancient Egypt. There were more than a

ABOVE: Copper-mining at Neusohl in the Carpathians. Woodcut from Georgius Agricola's *De Re Metallica*, Basel: 1557, Book VIII, University of California Libraries.

hundred Egyptian mines in the Nubian desert—the very name Nubia means "land of gold"—operating from around 2000 BC and worked by slaves. The precious metal adorned the pharaohs, and looked as bright as ever when the artifacts were recovered from their tombs. Much of the gold of the Roman Empire, meanwhile, came from mines at Rio Tinto,

LEFT: Solid gold statuette of the Egyptian God Amun-Re, possibly Thebes, Karnak, Middle Kingdom, ca. 945–712 BC, The Metropolitan Museum of Art, New York.

in Spain, which had been worked since around 1000 BC by Phoenician settlers and were also a source of copper and silver.

While silver does not have gold's cachet, it also impelled a great deal of industrial mining activity purely to access it as a repository of status. Silver often occurs in lead deposits as an impurity in the lead ore, galena (lead sulfide), from which it was extracted; sometimes veins of native silver also run through galena seams. Silver and lead were smelted together as an alloy from galena, and were separated in a process called cupellation, which was introduced around 3000–2500 BC. This involved melting the alloy in a clay crucible and blowing air over it to remove the lead through its reaction with oxygen, leaving a shining button of silver. Cupellation was later used to remove impurities, including silver, from gold.

A lust for gold has driven both scientific progress and world history. It was a key motivation for alchemical experiments that, in futile attempts to make gold from less valuable metals, resulted in all manner of other useful chemical discoveries. And it was gold that drew the Spanish conquerors, and then settlers, to the New World, and which supplied the impetus for the expansion of the European settlers in North America toward the West Coast in the nineteenth century. Yet, aside from some modern niche applications (as well as being the coloring agent for ruby-red glass in the Renaissance), gold has historically never been very useful. Until modern times, much the same might be said of silver. They are rare examples of elements long valued and enjoyed for the esthetic appeal of the pure material itself.

OPPOSITE: Vivid, ruby-red glass of The Lycurgus Cup, a *diatretum* (cage cup) made from dichroic glass, gilt, and silverwork, Late Roman Empire, fourth century AD, The British Museum, London.

TIN AND LEAD

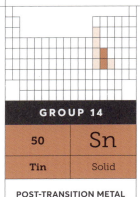

GROUP 14	
50	Sn
Tin	Solid

POST-TRANSITION METAL

ATOMIC WEIGHT: 118.71

GROUP 14	
82	Pb
Lead	Solid

POST-TRANSITION METAL

ATOMIC WEIGHT: 207.2

The Bronze Age tends to be associated with copper, but truly it was as much the Tin Age too. Since bronze is an alloy of those two metals, their early histories are inseparably entwined. Copper and tin ores often occur together, and bronze was sometimes made by smelting a mixture of these and letting the two molten metals combine in the furnace. It's possible that an accidental process like this led to the very discovery of bronze, although eventually the ores were deliberately mixed in carefully weighed proportions to make the desired kind of bronze alloy.

Tin is itself quite easily extracted from its main ore—a mineral called cassiterite, which is a reddish brown form of tin oxide. It is named after the Greek word (*kassiteros*) for the metal itself; the Romans called it *stannum*, from which tin takes its chemical symbol Sn. This became the French *etain* and the German *Zinn*, from both of which it is just a short step to the modern English word.

Tin smelting was happening in Europe by at least 1500 BC. There were tin mines throughout Europe; those in the southwest of England (Devon and Cornwall) were active from the early Bronze Age, around 2150 BC, and some historians believe that the British Isles might have been the "tin islands" (Cassiterides) first mentioned by the Greek writer Herodotus in the fifth century BC, to which the Phoenicians might have voyaged.

RIGHT: Cornish tin ingots, found off the coast of Israel, ca. 1300–1200 BC, courtesy of Ehud Galili.

LEFT: Romano-Egyptian mummy figure painted on a cedar of Lebanon wood panel using lead white, Egyptian blue, copper mineral green, hematite, and red, yellow, and brown iron oxides, Egypt, AD 220–235, The J. Paul Getty Museum, Los Angeles.

Tin is a relatively soft, silvery metal that can be hammered into sheets. The thinnest of these was tin leaf or "tin foil," which was used to cover objects and give them the bright appearance of silver—or, when glazed with translucent yellow pigment, as a cheap imitation of gold. (Today's "tinfoil" is made from aluminum instead.) Tin's resistance to tarnishing also made it a popular metal for making cooking and eating implements: pots and pans, tankards and teapots, not to mention containers for preserving canned food.

As we saw earlier, lead has a history bound up with silver, just as tin does with copper: the main ore of lead, galena (lead sulfide), has been mined for the

ABOVE: Corinthian ceramic *pinax* (plaque) showing miners at work in a quarry, 630–610 BC, Antikensammlung, State Museum of Berlin.

metal for several millennia. In fact, since lead can be smelted from this and other ores simply by heating it on a wood or coal fire, the practice may have taken place as early as 7000–6500 BC—lead beads of this age, for example, have been found in central Anatolia. There are also molded lead figurines from ancient Egypt from around 4000–3500 BC and lead coins from China and Assyria from 3000–2000 BC.

Lead has arguably the most unjustly gloomy reputation of all the "classical" metals. It seems to represent all that is ponderous, sluggish, and dirty:

too soft to be of much use for making tools or other implements, and poisonous to boot. It is the lowliest metal in the alchemical pantheon: the mere starting material in the quest to elevate other metals to gold.

Yet from lead came some of the most radiant substances in the ancient artist's palette. By corroding lead with vinegar fumes, the Egyptians made a white pigment, lead acetate (called lead white), which remained the painter's finest white right up until its replacement with zinc white in the nineteenth century. By heating lead in air, the ancient artisan could make the rich orangish red that the Romans called *minium* (red lead), from which the word miniature—a painting (often finely detailed) that makes abundant use of red lead—is

derived. Another form of lead oxide was bright yellow, and was used as the pigment massicot in the Middle Ages. The very softness of lead, as well as its relative abundance, made it a useful material for making water pipes and channels: "plumbing" comes from the Latin word for lead, *plumbum*, as does its chemical symbol Pb. The Romans made lead sheets by pouring a shallow layer of the molten metal into troughs of sand or earth and then bending and beating them into shape. Lead sheeting was also handy as a waterproof sealant, being widely used to protect church roofs against the elements.

A deadly reputation

Lead is, however, poisonous, and lead mining was deeply hazardous. The silver mines of Laurion, near Athens, in use from around 3200 BC, were also a key source of lead for the Greek city-state, and they rather belie its famous reputation as the fount of democracy. The miners were mostly slaves, some of them children, and many worked chained and naked. By the end of the first century AD some of the shafts were 350 feet deep. There is evidence of lead pollution from mining activities in Roman times.

The Romans understood the toxic influence of lead: the engineer Marcus Vitruvius in the first century BC noted that lead smelters were often pale and wan. All the same, the Romans produced a food sweetener called *sapa* (which is lead acetate and was later called sugar of lead) by boiling grape juice or old wine in lead pans. Some of the bones of people who lived in Roman London (Londinium) had lead levels more than seventy times higher than those of people living in pre-Roman Iron Age Britain. Elevated lead levels have also been found in the tooth enamel of citizens of Roman Europe. It's not clear why they had a greater lead exposure—perhaps *sapa* and plumbing played a part—but historians have suggested (speculatively, it must be said) that the problems caused by lead poisoning might have contributed to the decline of the Roman Empire.

BELOW: Inscription on a Roman lead tank, fourth century AD, from Suffolk, England, The British Museum, London.

IRON

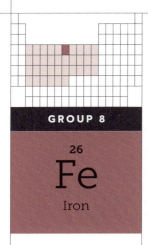

GROUP 8

26
Fe
Iron

Transition metal

ATOMIC NUMBER
26

ATOMIC WEIGHT
55.845

PHASE AT STP
Solid

Perhaps no discovery of an element has had such a transformational impact on world history than the smelting of iron from its ore. We can't be sure exactly when this happened, let alone how. But it seems to have begun in the Hittite Empire of Asia Minor in the thirteenth century BC, and the strong, tough iron weapons of the Hittite army made them almost impossible to resist with more brittle bronze armaments.

The process for making iron that could hold a sharp edge—which means, in effect, adjusting the amount of carbon mixed in with the metal to 0.1 percent or so—was refined by the Hittites from about 1400 BC, but it wasn't until their empire broke up and their metallurgical expertise was scattered far and wide that the Iron Age truly began, around 1200 BC.

This was not exactly the "discovery" of iron, however. Native iron can be found in nature, but very rarely: it is present in some meteorites (although the high temperatures needed to melt iron—around 2,795°F—would have made the metal of little practical use to any who stumbled over such a bounty in the Stone or Bronze Ages). And there are iron artifacts—ornaments and ceremonial weapons—dating from before 2000 BC. The crude "wrought" iron in these items, though, is no match for steel. The Hittites made steel by a process called cementation, which means hammering hot iron with charcoal to get carbon into it. The steel is hardened further by quenching—plunging the forged metal into cold water. These techniques weren't perfected until the first millennium BC, which is when the Iron Age took off. From around the ninth century BC, the metallurgical art of the Hittites was adopted by the

RIGHT: Assyrians besieging a city with an iron-tipped battering ram and iron weaponry. From the bronze gates of Shalmaneser III, at Balawat, ca. 865 BC, The British Museum, London.

水排

ABOVE: Earliest known illustration of a waterwheel-operated blast furnace, based on Du Shi's invention. From Wang Zhen's *Nong Shu* (Book of Agriculture), Yuan Dynasty, 1313, Vol. 6.

Assyrians who, in besieging Jerusalem in 701 BC, "came down like the wolf on the fold," as Lord Byron put it, "and the sheen of their spears was like stars on the sea."

According to the historians of technology Thomas Derry and Trevor Williams, writing in 1960, "the Greek civilization of the sixth century BC was founded upon iron, while the spread of Roman power, which eventually carried that civilization to the farthest limits of the Western world, was associated with iron throughout its long history." It was partly to gain access to iron mines that the Romans were keen to conquer parts of Europe, such as Spain: the Rio Tinto mines of Huelva province sit on a seam of copper-rich pyrite (iron sulfide), which is also known as fool's gold, and the iron-rich red soil of the region gives it its name.

The coming of steel

Smelting iron ore begins by roasting the mineral—pyrite, say—to make the oxide. Then the oxygen is removed by heating with carbon (charcoal), with which it combines to make carbon dioxide gas. In chemical terms this is a reaction called reduction, and it turns the iron present in the compound into its elemental, metallic form—which pools and can be drained from the kiln or furnace in its molten state. Early smelting methods didn't actually melt the iron, however, but produced a spongy mass called a bloom that could be hammered into wrought iron.

To make molten iron for casting, the temperature of the smelting process needed to be raised by blowing air through it, in a blast furnace. This process was used in ancient China from at least the first century AD, where its invention is attributed to Du Shi, an engineer and administrator of the Han Dynasty. It seems that some Chinese metallurgists

were already using hand bellows to smelt iron ore, but Du Shi showed how they could be operated automatically with a water wheel. Blast furnaces powered by water didn't become commonplace in Europe until the early sixteenth century, when the quality of Western iron and steel manufacture finally began to catch up with that of the East. Water power was also used to drive the hammers and rolling mills used to process and fashion the product, and by 1700 the iron industry had already largely undergone its "industrial revolution."

It wasn't until 1722, however, that the French polymath René-Antoine Ferchault de Réaumur showed that the properties of iron depend crucially on the amount of carbon it contains. Cast iron has the most, wrought iron the least, and steel finds an ideal value in between. De Réaumur did not quite express it this way, still being influenced by an alchemical way of thinking about chemistry: he said that it was the amount of "salts and sulfurs" in the iron that mattered. But he showed that, of the additives used to treat wrought iron, only those that contained carbon produced good steel. The role of such "extra ingredients" in steel was also carefully investigated half a century later by the Swedish chemist Torbern Bergman. Yet he too was limited by the chemistry of his age, expressing his conclusions in terms of the amounts of "phlogiston" and "caloric matter" in the iron: two fictitious elements that we

will encounter later. It wasn't until 1786 that three French scientists first expressed the matter clearly: "Cement steel is nothing other than iron…combined …with a certain proportion of natural charcoal [French: *charbone*]."

Once this was properly understood, the production of good-quality steel became more reliable, particularly after the introduction in the 1850s of Henry Bessemer's method for removing excess carbon (and other impurities) from molten iron by blowing air over it. Bessemer described his process in 1856, and patented it later that same year.

ABOVE: Using molds in an iron furnace. From René-Antoine Ferchault de Réaumur's *L'Art de Convertir le Fer Forgé en Acie*, Paris: Michel Brunet, 1722, Plate 23, University of Seville Library.

RIGHT: A Bessemer converter in action, blowing iron into steel, 1895. Underwood & Underwood Stereograph, the Science History Institute, Philadelphia.

OPPOSITE: "End of track, on Humboldt Plains." Photograph by A. Hart showing the iron railroad construction by Chinese workers on the Central Pacific Railroad, Nevada, 1865–1869, Library of Congress Prints & Photographs Division, Washington, DC.

BELOW: William Kelly's steel-making patent, 1857, United States Patent and Trademark Office.

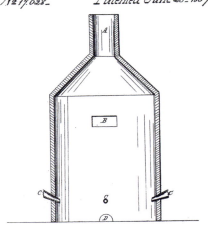

But his priority was disputed by an American named William Kelly, who developed much the same method in the early years of that decade. Kelly was convinced that his steel-making process had been mentioned in England and then copied by Bessemer. Kelly was granted a US patent in 1857, but it did him little good: he went bankrupt later that year and was forced to sell it, and it is with Bessemer's name that the invention is now generally associated.

Steel railway tracks were much longer-lasting than those made of wrought iron, and the railway networks, made now from Bessemer steel, began to grow apace from the late 1860s. By the end of the nineteenth century, steel was replacing wrought iron throughout the construction and transport industries: the modern Iron Age—more properly the Steel Age—was now upon us.

CHAPTER THREE

ALCHEMICAL ELEMENTS

LEFT: An alchemist carrying out a (symbolically depicted) chemical transformation. From Edward Kelly's *Theatrum Astronomiae Terrestris*, 1750 (Latin, Greek, and German manuscript), Saxon State and University Library, Dresden.

ALCHEMICAL ELEMENTS

Between the Middle Ages and the Renaissance, when chemistry meant alchemy, and the emergence of a recognizably modern form of this branch of science in the eighteenth century, it was often called chymistry. As the word suggests, this was a science in transition, being neither one thing nor the other, but in some ways a mixture of both. Of course, that's just the way it looks in retrospect. At the time, the words alchemy, chymistry, and chemistry were used interchangeably, and the chymists of the sixteenth and seventeenth centuries were doing what natural philosophers and scientists have always done: trying to figure out how the world works and how we can make use of that knowledge, hoping little by little to improve on the ideas that came before—but also, more often than was wise, asserting that they now had the answers which eluded their predecessors and peers. Chymistry was work in progress, as science always is.

Much of chemistry at this time was directed toward the vital task of making medicines. That chemistry could have this purpose was an old idea: much of the work of the Chinese alchemists from at least the Han Dynasty (202 BC—AD 220) was concerned with making health-giving elixirs. In the Middle Ages, chemical cure-alls called theriac, derived from ancient Greek and Roman recipes and containing ingredients such as roasted snake, could be bought in any self-respecting apothecary's shop. The use of chemical processes such as distillation to produce sophisticated chemical remedies was advocated in the fourteenth century by the Catalan physician Arnald of Villanova and the Frenchman Jean of Rupescissa. Their medicines might not have been effective, but their investigations helped to develop and popularize new chemical processes and substances: Arnald, for example, prepared almost pure alcohol by distillation.

Both of these figures influenced the sixteenth-century Swiss physician Paracelsus (his name, like many in the Renaissance, was self-styled in Latin; his real name was Philippus Aureolus Theophrastus Bombastus von Hohenheim, from a Swabian family of poor nobles). Paracelsus arguably did more than anyone in the Renaissance to turn alchemy away from a quest to make gold (although he tried that too) and toward medicine. One of Paracelsus's most celebrated remedies was laudanum, which was said to have miraculous properties: his assistant later claimed that with laudanum pills Paracelsus "could wake up the dead." No one knows what was in this alleged wonder drug, but a potion with the same name was promoted by the seventeenth-century English physician Thomas Sydenham; it was essentially opium dissolved in alcohol and seasoned with spices. It wouldn't have cured anything, but it would have eased a patient's aches and pains.

Medicine remained a strong theme in the chymistry of the century and a half after Paracelsus's death in 1541. But his vision was wider than the merely practical. He was a central figure in establishing a "chemical philosophy" that amounted to little short of an alchemical Theory of Everything. In his view, all that happened in the cosmos could be interpreted in chemical terms. The evaporation of water from

the oceans and its falling as rain, for example, mirrored the process of distillation in the alchemical laboratory. The human body was also governed by chemical principles: we all, Paracelsus said, have a kind of internal alchemist who transforms food into flesh, blood, and bone (which is, after a fashion, quite true). Even the biblical genesis, in which the land is separated from the water out of a primal Chaos, could be regarded as a chemical process.

This was, of course, a picture that owed as much to mysticism as to what we'd now regard as a scientific view of nature. But it's not fanciful to see within it the first glimpse of the idea that nature may be understood as a rational process which we can study in the lab—not so far, really, from the modern view of cosmology in terms of the Big Bang that physics can explain and investigate in particle accelerators. And there's no denying that it was a rather beautiful idea, which for a brief period placed chemistry and the chemical elements at the heart of things.

SULFUR

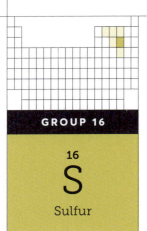

ABOVE RIGHT:
Rosicrucian portrait of Paracelsus (Philippus Aureolus Theophrastus Bombastus von Hohenheim) holding the sword said to have his Azoth or "universal cure" hidden in its pommel. Frontispiece to his *Philosophiae Magnae*, 1567, Science History Institute, Philadelphia.

GROUP 16

16
S
Sulfur

Non-metal

ATOMIC NUMBER
16

ATOMIC WEIGHT
32.06

PHASE AT STP
Solid

There has always been a whiff of devilry about sulfur, or "brimstone," as it was once known. That's not surprising, for natural deposits of sulfur often occur in rather hellish places: around volcanoes. Investigating a volcanic crater lake in Costa Rica in 1989, two British scientists discovered that the water had boiled away to reveal steaming pits of molten sulfur, crusted with bright yellow crystals of this element and stinking of the pungent gas—sulfur dioxide—which is formed when sulfur combines with air.

Since the pure element occurs in mineral form in nature, sulfur is one of those elements that did not need to be discovered. But it has always had uses, which is why sulfur deposits have been mined in volcanic regions since ancient times. Its very pungency made it a useful fumigant: burning sulfur to make sulfur dioxide gas was a means of driving away vermin such as mice, cockroaches, and fleas, and powdered sulfur would sometimes be sprinkled in food stores to keep such pests at bay. It was also used as a medicine: physicians thought it could help restore the balance of the humours, the four bodily fluids that were believed in ancient and medieval times to govern health. Arabic alchemists mentioned ointments that contain sulfur, and the influential Swiss physician Paracelsus and his followers recommended such treatments to cure itches.

Sulfur is inflammable, whence the association of brimstone with fire: the biblical book of Genesis records how "the Lord rained down burning sulfur on Sodom and Gomorrah" to punish the citizens of those places for their sinful ways. In Book II of *Paradise Lost*, John Milton describes Satan's realm as a place of stinking fumes, his very throne made from "Tartarean Sulphur and strange fire." And it could be hellish stuff, for sure. It was probably used as a component of the incendiary weapon called Greek fire, with which the Byzantine Empire waged naval warfare from around the seventh century AD. No one knows quite what went into this lethal mixture—and probably recipes varied—but most seem to have included sulfur along with inflammable substances derived from crude oil or resins. By all accounts it was almost impossible to extinguish, even while it floated on water.

Later, sulfur became an ingredient in gunpowder. This was famously invented in China, probably around the ninth century. It's sometimes said that the Chinese used it only for entertainment—for making the firecrackers

ABOVE: Depiction of raining sulfur in John Martin's *The Destruction of Sodom and Gomorrah*, 1852, Laing Art Gallery, Newcastle-upon-Tyne.

RIGHT: Islamic alchemist Jabir ibn Hayyan, possibly by Giovanni Bellini. From *Miscellanea d'Alchimia*, 1460–1475, Cod. Ashburnham 1166, Laurentian Medicean Library, Florence.

of which the Chinese people are still so fond—until its secret was leaked to the West about 250 years later, where it was promptly put to more deadly uses. But this isn't really true: it was being used in warfare in China from at least the eleventh century, for example in "fire arrows" and bombs flung at the enemy in sieges and naval battles. Gunpowder combines sulfur with charcoal and the compound known as saltpeter (potassium nitrate): the saltpeter supplies the oxygen that lets the sulfur and charcoal burn with such sudden fury. It's mostly the charcoal that causes the fiery explosion, but the sulfur helps it ignite at a lower temperature.

Sulfur was always of great interest to alchemists, who suspected it might be needed to make gold from other metals. The Arabic alchemist Jabir ibn Hayyan

held the notion that all metals are composed of the two "principles" sulfur and mercury, and that making gold was a matter of combining these in the right proportions. Paracelsus broadened this "unified theory of metals" to include all substances, by adding a third principle, salt. He argued that mercury was the principle that made things fluid, salt gave them "body" and made them solid, while sulfur was the principle of flammability: it made things burn.

The stench of brimstone

Some of the most influential texts in early alchemy were written around the late third century AD by Zosimos of Panopolis, a city in Roman Egypt. (At least, the texts used by later alchemists had his name attached, but we know very little about him or whether he really wrote them. Many works in

alchemy became attributed to famous writers as a way of making them sound more reliable.) Zosimos wrote about a substance called "sulfur water," which could be used to treat base metals so that they looked like gold. That involved a complicated process with many steps, each accompanied by a change in color of the metal—and it's likely that every step did involve some sort of chemical reaction, though it's not always easy to figure out what this might have been. Sulfur water was apparently applied to an alloy of lead, tin, copper, and iron to give it the yellowish hue of gold. Sulfur water itself was made by heating sulfur and lime (calcium carbonate) and dissolving the stuff that results—it seems to have been a solution of the gas hydrogen sulfide, which smells like rotten eggs.

That's the fate of sulfur: it's probably the biggest culprit in securing chemistry's reputation for bad smells. As well as acrid sulfur dioxide and rotten-odorous hydrogen sulfide, there are also sulfur-containing compounds called mercaptans, which vary in their smell from garlicky to the totally putrid aroma of rank boiled cabbage. Sulfur is also linked to flatulence, with sulfur-containing Brussels sprouts,

BELOW: Gunpowder fireballs being used by invading Mongol warriors. Detail from Takezaki Suenaga's *Scrolls of the Mongol Invasions of Japan*, 1275–1293 (paper handscroll with ink and colors), Imperial Household Collection, Tokyo.

Das anderte Buch

for example, getting both their distinctive (and to some, unappealing) smell and bitterness from molecules called glucosinolates, which, as the name suggests, are a little like sugars except that sulfur has muscled in on the act, to replace sweetness with quite the opposite. It's because of this quirk of its chemistry—or more properly, you might say, the way our bodies have evolved to respond to sulfur—that it will probably never shake off its satanic reputation.

ABOVE: The "alchemist's trinity" of sulfur, mercury, and salt. An allegorical image from Zoroaster's *Clavis Artis*, Ms-2-27, Vol. 3, 1858, Attilio Hortis Civic Library, Trieste.

RIGHT: Pottery caltrops (gunpowder-filled missiles), possibly Yuan Dynasty (1206–1368), National Museum of China, Beijing.

PHOSPHORUS

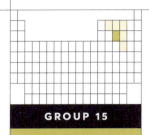

GROUP 15

15
P
Phosphorus

Non-metal

ATOMIC NUMBER
15

ATOMIC WEIGHT
30.974

PHASE AT STP
Solid

RIGHT: Making the element phosphorus with urine. Detail from Robert Boyle's *Way of Making Phosphorus*, 1680, The Royal Society, London.

Probably no story of the discovery of an element beats that of phosphorus. It has everything: drama, intrigue, mystery, hardship, excitement, danger—and bad smells. More than any other of these tales, it illustrates the value in scientific discovery of following your nose—in this case, quite literally.

Hennig Brandt was an alchemist at a time when alchemy was fading—or rather, was changing into the discipline we now call chemistry. He worked in Hamburg in the mid-seventeenth century. Little is known about his background except that he was also a glassmaker and believed in the philosopher's stone, the substance that could allegedly transform base metals into gold. The wealth that this agent of transformation promised had enticed alchemists for centuries, and Brandt had good reason to fall for that allure. His laboratory was financed by the dowry from his first wife and, after she died, by the resources of the wealthy widow he married subsequently. These financial means weren't enough, however, and Brandt was always on the lookout for turning a profit from his alchemical researches. He came by the idea—which seems bizarre in retrospect—that the key ingredient for the philosopher's stone might be distilled from urine. From around 1669, he began to gather quantities of urine and distill them to extract the solid residue.

Brandt found that there was indeed a substance left in the flask, which

OPPOSITE: Joseph Wright of Derby's *The Alchymist, In Search of the Philosopher's Stone*, 1771, Derby Museum and Art Gallery, England.

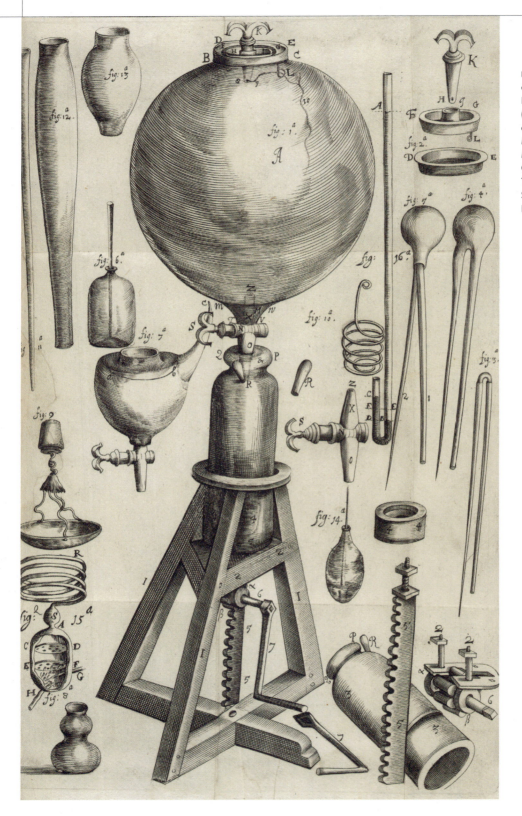

LEFT: Copperplate engraving of Robert Boyle's air-pump (pneumatical engine) and components. From *New Experiments Physico-Mechanicall*, Oxford: H. Hall for T. Robinson, 1660, Science History Institute, Philadelphia.

melted when heated to produce a garlic-smelling liquid that glowed with its own light and burst into flame when in contact with air. He collected this soft solid and kept it secret for six years as he tried to turn it into the philosopher's stone. His discovery was immortalized a century later by the English painter Joseph Wright of Derby, who shows the alchemist kneeling in a laboratory that looks like a medieval Gothic crypt—he could be a monk being granted some divine revelation. From the flask in front of him comes a radiance that floods the scene and creates dramatic shadows. Wright of Derby was wanting to draw parallels between the religious experience of revelation and the wonder of scientific discovery: it is an allegory for the process of "enlightenment" which was then thought to herald the dawn of the age of science.

Despite his secrecy, news of Brandt's discovery leaked out. On hearing of it in the mid-1670s, Johann Kunckel, a professor of chemistry at the University of Wittenburg, decided to track Brandt down. But he wasn't the only one. Kunckel had written, to his later regret, to a colleague in Dresden named Daniel Krafft, who figured that the story might be worth investigating. The tale goes that Krafft got there first and was in the midst of negotiating a price with Brandt for his supply of the glowing substance when Kunckel turned up and implored to be told the secret of how to make it. Brandt seems to have revealed only that it came from urine—which was enough for Kunckel to start distilling urine himself. He succeeded in making the stuff in 1676.

Krafft was already peddling his pot of the new material. It became known as "phosphorus"—literally, bearer of light. No one knew that it was a new element in its own right, and the word was used indiscriminately in the seventeenth century for any substance that spontaneously glowed, a property called phosphorescence. Krafft had been taking Brandt's phosphorus around the European courts, charging a princely sum for a demonstration of its properties. His display is described in an account by the Anglo-Irish scientist Robert Boyle (see page 72) for the new Royal Society of London, a group of natural philosophers interested in all things novel.

Krafft visited Boyle's home, Ranelagh House, in London, in September 1677, with various vials, tubes, and flasks of solids and liquids. One of these, said Boyle, contained a reddish liquid that shone "like a cannon bullet taken red hot out of the fire." Krafft dipped his finger into some of the phosphorus and wrote the glowing word DOMINI; he scattered pieces of the material on the fine carpets of Boyle's sister (whose house this actually was), where they glittered like stars.

Boyle, a curious man and an accomplished chemist, was desperate to know how to make this stuff himself, but all Krafft would tell him was that it came from within "the body of man." He correctly guessed that the source was urine and hired a German assistant named Ambrose Godfrey Hanckwitz to help him make it. Godfrey (as he was known) traveled to Hamburg to find out more about that process from Brandt himself, whose role in the discovery only became clear after he was found again by the German philosopher Gottfried Leibniz, who wrote about it in a letter to the Royal Society. Godfrey succeeded in distilling phosphorus within his lodgings in London, supplying Boyle with as much of the enchanting element as he wanted.

Phosphorus is odd stuff. It exists in several pure forms: red, white, and black. White phosphorus is the most common, and is soft and waxy, melting at just 111°F. It glows when exposed to oxygen due to chemical reactions between the two elements that cause the phenomenon of chemiluminescence: they make compounds that emit light. It is also apt to burst spontaneously into flame in air, and so is used in incendiary weapons and as "tracer ammunition" to show the path taken by other projectiles.

It is awful stuff really: it can cause serious burns if it comes into contact with the skin, and it is also highly toxic. And yet—again we see the paradox of chemistry—phosphorus is an essential element in living systems. Combined with oxygen in the form of phosphate, it makes up the fabric of bone and the links in the backbone of the DNA molecule, the bearer of genes in all organisms. It is because there is so much phosphorus in our bodies that there's also plenty in urine, in which any excess is excreted. It is one of the most marvelous and perplexing of all the elements in nature.

ANTIMONY

GROUP 15

51

Sb

Antimony

Metalloid

ATOMIC NUMBER
51

ATOMIC WEIGHT
121.76

PHASE AT STP
Solid

OPPOSITE: Frontispiece to Theodor Kerckring's commentary on Basil Valentine's *The Triumphal Chariot of Antimony*, 1624, Amsterdam: Andrea Frisii, 1671, The Getty Research Institute, Los Angeles.

Many elements were first recognized and named only in the form of compounds: in combination with other elements. Antimony is one of those. It occurs in nature in mineral form as its sulfide, which the Romans called *stibium* and which today we call stibnite. Therein lies the origin of the (otherwise perplexing) modern chemical symbol Sb.

Stibnite is soft and black, and it was used in powdered form, mixed with waxes or oils, as a cosmetic by the Egyptians. The Assyrian word for "eye paint," *guhlu*, is the root of the Arabic *kuhl*, and even today black eye cosmetics such as mascara are sometimes called kohl. Stibnite ointment was also thought to cure eye infections, and Dioscorides, a Greek physician of the first century AD, listed it as a drug.

This substance was long thought to have healing properties. In the thirteenth and fourteenth centuries, two Catalonian alchemists known as Arnald of Villanova and John of Rupescissa used distillation to produce a wide variety of new substances which they called spirits or quintessences—a remnant of Aristotle's word for the heavenly fifth element or aether. One of these was made from stibnite, dissolved in vinegar and distilled, and John of Rupescissa claimed that it formed as "blood red drops" and "surpasses the sweetness of honey."

ABOVE: Woman with kohl (antimony-based makeup) holding a sistrum (instrument), Deir el-Medina, Thebes, New Kingdom, ca. 1250–1200 BC, Walters Art Museum, Baltimore.

Kohl is the root of the word "alcohol": in Arabic *al-kohl*. Isn't that odd?—that a term for a black mineral should come to designate a clear and volatile liquid? But that kind of slippage isn't uncommon in alchemy. At first, al-kohl meant powdered stibnite, then any kind of powder, and then any kind of "essence" of a substance, like the quintessences that alchemists had learnt to make by distillation. The word finally became attached to just one of these: the "spirit of wine."

"Antimony"—that is, compounds of this element, such as stibnite—was among the remedies that was favored by the influential physician Paracelsus, and his followers in the seventeenth century recommended it eagerly. These "Paracelsian" doctors were sometimes called iatrochemists—the word

CVRRVS
TRIVMPHALIS
ANTIMONII.

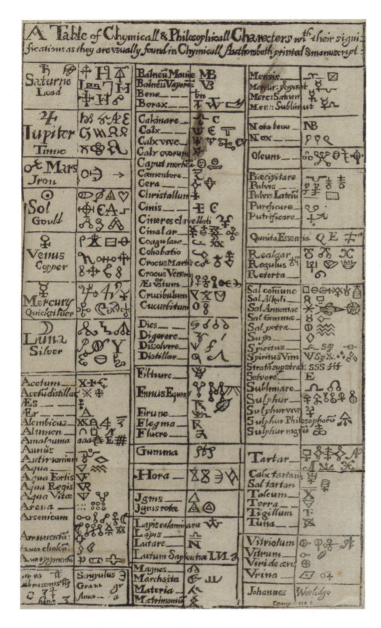

iatrochemistry means "medical chemistry"—and they opposed the old notion that health was all about balancing the four humours of the body. Instead, they insisted that specific ailments needed to be treated with specific chemical medicines, and that the doctor's task was to prepare and prescribe them. This idea motivated a lot of useful chemistry, but it didn't always lead to good cures. Antimony itself is fairly toxic: Mozart might have died from taking too much of the antimony salts his doctors prescribed, and it was also a favorite among Victorian poisoners. One seventeenth-century book called *The Triumphal Chariot of Antimony* claimed that the name is derived from *anti-monachos* (anti-monk or monk killer), because antimony cures were said to have poisoned a

La calcination Solaire de L'antimoine.

LEFT: Burning antimony with focused rays of sunlight. From Nicaise Le Fèvre's *Traité de la Chymie*, Paris: Thomas Jollie, 1669, Vol. 2, Science History Institute, Philadelphia.

community of Benedictine monks who took them. But the name probably (no one knows for sure) comes from *anti-monos*, meaning "not alone," perhaps because stibnite—the main antimony ore—was generally found alongside other minerals. The first recorded use of the word appears in an eleventh-century book by an Arabic alchemist.

The antimony wars

The Triumphal Chariot was a salvo in a fierce argument that raged between the Paracelsians and the traditional physicians, especially in France, over the value of antimony medicines; this has even been dubbed the "Antimony Wars." The Paracelsians argued that even though they knew antimony could be toxic, their chemical processes separated the good influences from the bad. These disputes were really, however, all about authority: the Paracelsians and the traditionalists were rival camps vying for influence in the French court, and who was correct about antimony really represented a struggle to assert preeminence in all of medicine. It was also a battle for the soul of chemistry itself: could it be entrusted to Paracelsians, who used arcane terms and harked back to the disreputable age of alchemy,

or was it time for chemistry to become a transparent and rational science?

It's not known when pure elemental antimony was first separated from its ore. But this is not so hard to do—simply heating the sulfide in air will drive off the sulfur—and it was even done in ancient times. There have been claims of objects containing antimony metal from the Middle East and Egypt dating from the third millennium BC, but that's disputed by archeologists. The pure element looks like a grayish metal, similar to lead. It's not a true metal, though—it is what is known as a metalloid, which doesn't conduct electricity as well as proper metals do. Still, it can be mixed with other metals, such as lead and tin, to make alloys. One of these was used to make printing dies, because it expands slightly as it cools and solidifies, and can therefore be cast into sharp-edged objects.

The toxicity means the metal causes intestinal upset: it acts as a laxative. People used small pellets of pure antimony to cure constipation in the Middle Ages. These pills weren't cheap, and so they would be carefully retrieved once they had done their job and been expelled, in order to be used again and again. It's best not to think too hard about that.

PHLOGISTON

When Paracelsus stated that sulfur was the "principle of combustion," he established the idea that everything that burns does so because of an ingredient it contains. Paracelsus's "principles" weren't like our modern idea of elements—they weren't exactly substances that you could isolate and purify, though historians of chemistry still argue about whether the "sulfur" and "mercury" thought to be the constituents of all metals were considered to be the same as the yellow mineral and silvery liquid metal that alchemists could make. Little by little, though, these "principles" did evolve into something more like elements in the conventional sense. The principle of combustibility became, in the eighteenth century, probably the most famous "element that never was," which early chemists called phlogiston from the Greek word for "to set on fire."

That was a gradual process. First, the German alchemist Johann Joachim Becher modified the Paracelsian scheme in claiming that there were three types of "earth": a fluid sort (like mercury), a solid sort (like salt), and a fatty and combustible sort (like sulfur). Becher was something of a throwback: an alchemist of questionable repute who toured Europe in the late seventeenth century persuading people to give him money in return for making gold. His ideas, if not his gold-making arts, convinced a chemist at the University of Halle named Georg Ernst Stahl, who in 1703 edited and published a new edition of Becher's great treatise on minerals in which his "fatty earth" was renamed phlogiston. When a substance burns, said Stahl, the phlogiston it contains is released into the air. This explains why, as wood burns, it gets lighter: the ash that remains is a mere fraction of the original material. Sulfur itself, said Stahl, is a mixture of "vitriol" (what we now call sulfuric acid) with phlogiston.

A burning question

Stahl's idea seemed to make a lot of sense, and phlogiston theory was believed by most chemists throughout the eighteenth century. They used it to explain not only combustion but also respiration, acids, alkalis, and more. They thought that air could become "saturated" with the phlogiston given off in burning, so that it could absorb no more—and that's when burning would cease. This was thought to be

ABOVE: Frontispiece to Johann Joachim Becher's *Physica Subterranea*, which introduced *terra pinguis* (fatty earth), Leipzig: Ex Officina Weidmanniana, 1738, University of Miskolc.

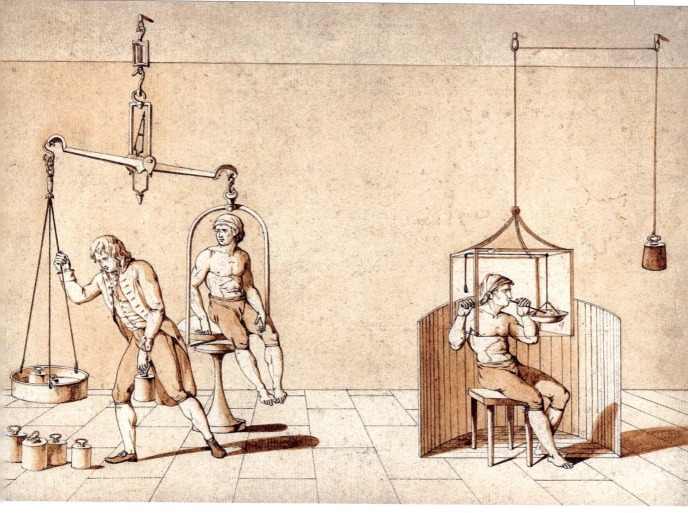

ABOVE: Marie-Anne-Pierrette Paulze Lavoisier's pen and wash drawing of her husband Antoine Lavoisier's experiments with respiration, ca. 1789, Wellcome Collection, London.

RIGHT: Apparatus for breathing dephlogisticated air. From Jan Ingenhousz's *Nouvelles Expériences et Observation Sur Divers Objets de Physique*, Paris: Chez P. T. Barrois le Jeune, 1789, Vol. 2, Plate IV, Wellcome Collection, London.

why a lighted candle would eventually go out when a bell jar was placed over it. Conversely, if air could be robbed of some of its phlogiston, it was better at sustaining combustion. When the eighteenth-century British scientist Joseph Priestley produced oxygen gas by heating mercury oxide and found it

LEFT: Wolfgang Philipp Kilian's engraving of the chemist Johann Joachim Becher, 1675, Wellcome Collection, London.

desperately, that maybe sometimes phlogiston could have "negative weight." It wasn't until the 1780s that phlogiston theory was shown to be wrong, mainly through the careful experiments of the French chemist Antoine Lavoisier. He proved that substances burn not by releasing some element (phlogiston), but by absorbing or combining with an element in the air—which he called oxygen. Metals gain weight when they "burn" because they combine to form the oxide. Lavoisier's theory was gradually accepted from around the end of the century—but grudgingly so by some, especially in England, where there was a chauvinistic resistance to "French ideas." Priestley went to his death in 1804 still doggedly clinging on to phlogiston.

So phlogiston was one of the most notoriously mistaken ideas about the chemical elements. It would be wrong to ridicule it, though. Phlogiston was an idea that helped chemists to bring some order to the diverse ways that substances behave, and this made it useful for the progress of chemistry. The problem wasn't that it was wrong, but that it was so nearly right: phlogiston theory was almost exactly the opposite of Lavoisier's oxygen theory of combustion. This made it relatively easy for Lavoisier's ideas to supplant it, without requiring lots of other changes to the way chemistry was conceptualized. It's actually an excellent example of why, sometimes in science, whether an idea is useful can matter as much as whether it is correct.

would make a burning cinder glow more brightly, he regarded the gas instead as "dephlogisticated air." Others who made hydrogen gas by reacting a metal with acid and found that it would ignite explosively suspected it might be pure phlogiston.

Phlogiston's opposite

Still, there were problems. For one thing, when metals were roasted in air, they didn't lose weight (by emitting phlogiston), but gained it. How could that be? Some chemists suggested, rather

OPPOSITE: Retorts and other apparatus. From Denis Diderot and Jean le Rond d'Alembert's *Encyclopédie, ou Dictionnaire Raisonné des Sciences, des Arts et des Métiers*, Paris: André le Breton et al, Avec approbriation et privilege du Roy, 1763, Part II, Plate XI, University of Chicago.

OVERLEAF: A chemical laboratory, with a table of the elements below. From Denis Diderot and Jean le Rond d'Alembert's *Encyclopédie, ou Dictionnaire Raisonné des Sciences, des Arts et des Métiers*, Paris: André le Breton et al, Avec approbriation et privilege du Roy, 1763, Part II, Plate I, Wellcome Collection, London.

fig. 146.

fig. 147.

fig. 148.

fig. 149.

fig. 150.

fig. 151.

fig. 152.

fig. 153.

fig. 155.

c

fig. 157.

fig. 156.

fig. 154.

fig. 162.

fig. 158.

a

fig. 159.

fig. 160.

fig. 161.

fig. 160. N.º 2.

WHAT EXACTLY IS
AN ELEMENT?

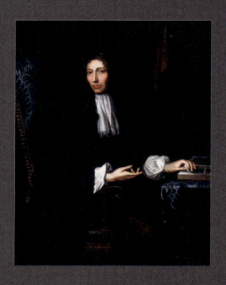

It has taken a lot of careful work by historians of science to start to repair alchemy's unwarranted image as pseudoscientific mysticism. Until relatively recently, it was portrayed as the practice of quacks, charlatans, and fools trying or claiming to be able to make gold—or, alternatively, as some kind of allegorical search for spiritual enlightenment. In fact, a great deal of alchemy was simply the practical craft of using chemistry to make useful substances such as pigments and dyes, soaps and medicines (although admittedly many of them weren't effective). If alchemists lacked any real understanding of what they were doing—or at least, did not express it in terms modern chemists would recognize—that was true of most science and technology before the seventeenth century. It was then that alchemy began to morph into "chymistry," the precursor of today's chemistry.

LEFT: Johann Kerseboom's *The Shannon Portrait of the Hon. Robert Boyle F. R. S.*, 1869, Science History Institute, Philadelphia.

This misconception about alchemy long obscured our understanding of what some of those pioneering "scientists" (the term hadn't yet been coined) in the seventeenth century were up to. The alchemical studies of Isaac Newton were deemed something of an aberration and embarrassment for so profound a thinker, and were glossed over. The same was true for Robert Boyle (1627–1691), the son of an Irish aristocrat raised in England, who ranks alongside Newton as one of the greatest natural philosophers of the late seventeenth century.

Boyle's most celebrated book, *The Sceptical Chymist* (1661), used to be seen as a broadside against the ignorance and deceptions of the alchemists. But it wasn't really that at all. Boyle believed in much of what alchemy stood for, including the possibility that base metals could be transmuted into gold, and he expended much effort trying to find the "philosopher's stone" that could allegedly accomplish this. But what he sought to do in his book was to draw a distinction between learned "chymical" science based on careful observation of experiments, and the mystics, dupes, deceivers, and vulgar recipe-followers who pretended to be able to do things they could not, or dressed up their ignorance in overblown language. "Believe me," he wrote, "when I declare that I distinguish betwixt those Chymists that are either Cheats, or but Laborants, and the true *Adepti*"—meaning knowledgeable scholars like himself. He had a point—there were plenty of dubious claims and practices in alchemy—but in the end this was not so different from what alchemists had *always* said in their disputes: you lot are ignorant "puffers" or charlatans, while I have *true* knowledge.

The Sceptical Chymist is itself not the most elegant or easily understood of Boyle's works. One of his main targets for attack is the idea, promoted by the followers of the Swiss alchemist and physician Paracelsus, that all matter is made up of three "principles": sulfur, salt, and mercury. It's not as simple as that, Boyle said—and there is no evidence that those three substances are truly *elements*, meaning the fundamental constituents

of matter. Neither, he said, can we fall back on the four classical elements of the Greeks: "Out of some bodies, four elements cannot be extracted." Take gold, he wrote—"out of which not so much as any *one* of them hath hitherto" been extracted. There are, Boyle said, probably more than four elements— but "no man had ever yet made any sensible trial to discover their number."

This raises the question: what, then, exactly *is* an element? And how do you know one when you see one? *The Sceptical Chymist* is often celebrated as the first book to provide something like the modern definition of a chemical element. Boyle said that they are,

"certain primitive and simple, or perfectly unmingled bodies; which not being made of any other bodies, or of one another, are the ingredients of which all those called perfectly mixt bodies are immediately compounded, and into which they are ultimately resolved."

In other words, you can't break down or separate an element into anything simpler.

Yet this is still a rather abstract and philosophical way to think about an element, and Boyle is silent about what such elements might be. In fact, he even wonders if anything like this, fundamental and irreducible, exists at all. His definition sounds similar to the one the French chemist Antoine Lavoisier offered in the late eighteenth century: a substance that can't be broken down by chemical reactions into anything simpler. But Lavoisier's definition was much more securely rooted in practical chemistry: he was a master at chemical *analysis*, which literally means splitting apart into the basic ingredients. For Boyle, an element remained a handy conceptual tool. He certainly didn't see any reason to consider gold to be "fundamental" in this sense, because he believed all his life that it could be made by "chymical" methods: a goal that was as alluring to him as it had been to alchemists for centuries.

It's a slightly odd thing, then, that of the several comprehensive and recognizably "modern" textbooks on chemistry written in the seventeenth

ABOVE: Colored lithograph of a man made out of chemical apparatus, symbols, and chemicals, early nineteenth century, Wellcome Collection, London.

century, Boyle's *The Sceptical Chymist* has come to be regarded as the exemplar, and as the one that revolutionized the concept of an element. Perhaps this is part of the tendency that used to exist in the history of science to search for foundational moments, figures, and texts. Until quite recently, Boyle's alchemical interests were (like Newton's) ignored or suppressed, and there was an eagerness to present the Royal Society, over which these two men presided, as setting the template for modern science. All the same, *The Sceptical Chymist* was a part of, if not the origin of, something new: a determination to be guided, in our concepts of elements and chemistry, by careful observation and experiment rather than preconceptions about how things ought to be.

CHAPTER FOUR

THE NEW METALS

LEFT: Mining ore. Depicted on a panel of Hans Hesse's *Annaberg Mountain Altar*, St. Ann's Church, Annaberg-Buchholz, Germany, 1522.

THE NEW METALS

M ining has been practiced since ancient times. But whereas back then its bounty served the might of states and empires, and the glorification of kings and emperors, in the late Middle Ages there were new beneficiaries: the merchant classes. With the fabulous wealth that some of them acquired came political clout to rival that of lords and rulers: power and authority were now no longer conferred by divine decree, but could be dug from the mineral riches of the earth. In such ways, the commerce in metals in the Western world transformed its entire social structure.

Take the Fuggers of Augsburg. Johann Fugger set up a cotton textiles business in that city in the 1360s. When it thrived, he diversified first into fine textiles like silk, then into imported spices. One of his sons acquired rights in a silver business in the Austrian Tyrol, and by the mid-fifteenth century the Fuggers were immensely rich. They started to loan money to the Archdukes of Tyrol, and it was in repayment for such a sum given in 1491 to Archduke Maximilian that Johann's grandson Jakob gained control over all the copper and silver mining in the Tyrol. When Maximilian became Holy

BELOW: Circular reverberatory furnace for smelting ores. From Vannoccio Biringuccio's *De La Pirotechnia*, Venice: C. Navò, 1540, Smithsonian Libraries, Washington, DC.

ABOVE: Albrecht Dürer's portrait of Johann Fugger "The Rich" (who built a mining empire that stretched across much of Europe), 1518, Staatsgalerie, Augsburg.

Roman Emperor in 1493 his bills escalated, and as he fell ever more in debt to the Fuggers they extended their mining empire across Europe, from Spain to Hungary. By the early sixteenth century they were one of the most influential bankers in Christendom.

Mining of metals was one of the most lucrative enterprises in the German lands. Lead and silver had been mined in the Harz Mountains of northern Germany from the tenth century, and silver deposits were discovered in 1136 in the hills between Saxony and Bohemia. The skill of German miners was famous throughout Europe, where the German tongue became the *lingua franca* of mining.

Mining and scientific advances

The scale of this enterprise is clear from the great 1556 treatise on mining, *De Re Metallica* (On the Nature of Metals), by the German Georgius Agricola. Here woodcuts show massive waterwheels for hauling minerals out of mineshafts and grinding them up; streams being diverted to drive the machinery and wash the ore; trees felled to provide timber and fuel; and workshops where the ore was sifted and smelted. Mineshafts could reach depths of 500 feet or more, requiring powerful machinery to pump water from the tunnels, and it's clear from Agricola's book that the profits from mining activities were costing nature dearly. He knew his audience, however, providing mineowners with arguments to defend themselves against charges of avarice and exploitation: the immense usefulness of metals, he says, far outweighs the nuisance and despoliation involved in extracting them.

The wealth to be had from mining stimulated efforts to understand minerals and their metal bounty. In this way, mining forged a link between speculative science and practical technology: both were a kind of alchemy. The Fuggers established mining schools (*Bergschule*) where teachers would instruct apprentices in the arts of metal-making. The commercial value of silver, copper, tin, and lead—the main metals to be had from the German lands—was clear enough, but metalworkers also began to appreciate that the ores might hold other

ABOVE: Scaling down into the pits. From Georgius Agricola's *De Re Metallica*, Basel, 1556, Book 6, Wellcome Collection, London.

metallic substances too, not recognized in the writings on mineralogy from antiquity. Some of these found their own profitable niche in the marketplace, as the demands of commerce motivated discoveries of new elements.

BISMUTH

GROUP 15

83
Bi
Bismuth

Post-transition metal

ATOMIC NUMBER
83

ATOMIC WEIGHT
208.98

PHASE AT STP
Solid

In ancient times, philosophers and craftspeople recognized seven different metals: gold, silver, mercury, copper, iron, tin, and lead. That was a neat scheme, because each metal could be assigned a heavenly body: the Sun, Moon, and the five known planets Mercury, Venus, Mars, Jupiter, and Saturn. Why do that, though? Because many natural philosophers believed that nature was governed by "correspondences" like this between different classes of things.

However, this neat idea got awkward when it came to alloys—mixtures of metals—like bronze and electrum. Were they different metals too? We've already seen that there was at least one other metal-like substance known since antiquity too: antimony, with the dull sheen of lead. But it wasn't alone. There was another, also dense and silver-gray like lead, but with a pinkish tint and more brittle, which became known as bismuth.

The origin of the name is a mystery. Some say it comes from the Arabic *bi ismid*, meaning "like antimony" (which it is); others that it is a form of the German *Wismuth*, meaning "white mass." Bronze objects containing bismuth exist from ancient times, although it was probably there by accident as a natural impurity within the tin used for the alloy. In a ceremonial bronze knife with around 18 percent bismuth found at the Inca city of Machu Picchu (which flourished in the late fifteenth century), however, it seems to have been added by design, maybe to make the metal easier to work.

Bismuth seems to have emerged in Europe around the early 1400s, although it wasn't always recognized as a metal in its own right, often being confused with lead, tin, antimony, or, later, with zinc. By the latter part of that century, however, there were metalworkers who specialized in working with bismuth—they even had their own guild. In the mid-sixteenth century, Georgius Agricola described how bismuth was mined and smelted from its ore. In one of his books, posed as a dialogue between an expert metallurgist called Bermannus and an apprentice appropriately named Naevius, the master tells his pupil of a metal "unknown to the Ancients" called *bismetum*. So you mean there are more metals than seven, asks the apprentice with surprise? More indeed, Bermannus attests.

The Spanish miner Álvaro Alonso Barba agreed with that view in his 1640 book *The Art of Metals*. In the mountains of Bohemia, he wrote, "was found a metal which they call *Bissamuto*, which is a metal between tin and lead, and yet distinct from them both." Yet by 1671, the Englishman John Webster wrote in his *Metallographia, or A History of Metals* that he couldn't get hold of any

OPPOSITE: Ores of bismuth, cobalt, arsenic, and nickel. From Louis Simonin's *La Vie Souterraine*, Paris: L. Hachette, 1869, Plate V, Science Information Center, University of Toronto.

ABOVE: Spa Well, Low Harrogate, Yorkshire, 1829. Lithograph by Day & Haye after J. Stubbs, Wellcome Collection, London.

bismuth in all the British lands. (To be honest, Webster couldn't have looked very hard, because by then bismuth had been produced in the copper and lead mines of England's Lake District for a hundred years.) It was often alloyed with tin in order to create a kind of hard pewter: some called this the "chemist's basilisk," an alchemical allusion to the way that mythical beast, a serpent, was said to turn people to stone by its gaze.

Bismuth was also used in cosmetics: it was reacted with nitric acid to make a white powder of the nitrate, the "Magistery of Bismuth," which was used as a face whitener to hide skin blemishes. The compound can be converted into black bismuth sulfide if exposed to hydrogen sulfide fumes—as one nineteenth-century lady allegedly found to her cost when, whitened by bismuth powder, she took a bath in the sulfurous spa waters of Harrogate and, emerging blackened, shrieked and fell into a swoon.

Despite all this, bismuth was not "officially" discovered until 1753, when the French chemist Claude François Geoffroy declared it a true element. If that seems confusing, it reflects the vagueness in those times about what truly constitutes an "element."

OPPOSITE: Collecting molten bismuth. Woodcut from Georgius Agricola's *De Re Metallica*, Basel, 1556, Book 9, The Getty Research Institute, Los Angeles.

ZINC

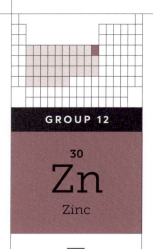

GROUP 12

30

Zn

Zinc

Transition metal

ATOMIC NUMBER
30

ATOMIC WEIGHT
65.38

PHASE AT STP
Solid

I n 1558, Agricola's narrator Bermannus mentions another mineral, which he says can be found in the region of Silesia (mostly in modern Poland) and which is called *zincum*. But he seems to think it is a kind of ore, not a metal.

Zinc ore generally occurs as the oxide of that metal, which was known in ancient times as *cadmia*, after the Greek hero Cadmus who allegedly founded the city of Thebes. The Roman writer Dioscorides in the first century AD says that a kind of cadmia also forms during copper smelting—copper ore often has a fair amount of zinc in it too. Dioscorides adds that cadmia can also be made by "burning a stone called pyrites" which is found in Cyprus (this too might have been a copper ore). He mentions other substances formed in the chimneys of copper furnaces, called *pompholyx*, *tutia*, and *spodos*—which are probably all zinc compounds. It all gets rather confusing—and to make matters worse, there was another common zinc ore, the sulfide, which is black and later became known by the nickname "black jack."

This is the difficulty with figuring out the metal elements: they often appear together in minerals, because chemically they can be rather similar, and so it was hard for ancient smelters and metallurgists to decide if what they were seeing was a different form of a known metal, or something genuinely new. Brass, for example, is an alloy of copper and zinc, being less red and more golden than copper itself—and it might have been formed by accident when copper was smelted from an ore rich in zinc. That's why brass was known long before zinc was recognized as a metal in its own right. Brass objects from parts of the Middle East and central Europe date back to the third millennium BC, and in China they appear even earlier. Brass-making was well developed in Rome, with Dioscorides explaining how to use cadmia from copper smelting to make it. The Romans used brass for coins such as the *dupondii* and the *sestertii*.

Who first made pure zinc metal isn't known—but zinc objects have been found in Greek ruins from around 500 BC and at some Roman sites too. The Greek writer Strabo, in the first century BC, refers to a "mock silver" which might be zinc. The first large-scale production of zinc metal happened in

RIGHT: Medieval zinc jittals from Kangra, Himachal Pradesh, North West India. Series B, C & M Reg. 1892, British Museum, London.

India from around the thirteenth century, and this technology found its way to China, where zinc was being produced in the sixteenth century.

The earliest clear reference to it in the West appears around the same time. In a book on minerals written by Paracelsus around 1518 (but not published until 1570), he says "there is another metal generally unknown called *zinken*. It is of a peculiar nature and origin…Its colour is different from other metals and does not resemble others in its growth." He called it the "bastard offspring of copper."

Metal of India

Paracelsus didn't, however, invent the word zinc, as has been sometimes suggested. Some fourteenth-century Spanish texts refer to *cinc* (still the modern Spanish word for zinc), although it appears there to refer to brass. Some think that the curious name stems from the Arabic *sini* and Persian *cini*, which refer to metals from China. At any rate, Europeans were importing it from India and the Far East by the late sixteenth century, and it was sometimes called "Indian tin." Shakespeare's mention of a "metal of India" in *Twelfth Night* might, in fact, be a reference to zinc. Zinc metal was prized because it was much better than cadmia (zinc oxide) for making brass: brighter and more gold-like in appearance. "Zink," wrote the German chemist Georg Stahl, who we met earlier as the inventor of phlogiston, "gives the Copper a much more beautiful colour than the Calamy [cadmia]"—making it into what became known as "Prince's Metal."

A German mining official gave a rather clear account of zinc in 1617, which made it plain that it differed from other metals. It has "a great resemblance to tin," he wrote, "but it is harder and less malleable, and rings like a small bell." Although made as a by-product of copper smelting, he says, "it is not much valued, and the servants and workmen only collect it when they are promised drink money." Still, he adds, when alloyed with tin it was much in demand by the alchemists.

Some confusion persisted. Robert Boyle talked about his experiments in 1673 on a metal called

ABOVE: Zinc oxide by-product being scraped off the walls at the Rammelsberg silver-smelting works. Woodcut from Lazarus Ercker's *Beschreibung Allerfürnemisten Mineralischen Ertzt unnd Bergkwercks Arten*, Prague: G. Schwartz, 1574, Wellcome Collection, London.

Tutenâg that was imported from the East Indies and was "unknown to the European Chymists," without realizing that it was the same as his familiar "Zink"—the odd word was related to the old Latin term for zinc oxide, *tutia*. Element-naming has always been a jumble, with old words reused in new guises and multiple names being given to the same substance. If, for example, you think that "cadmia" sounds familiar—well, wait and see.

COBALT

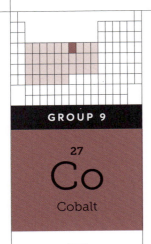

GROUP 9

27
Co
Cobalt

Transition metal

ATOMIC NUMBER
27

ATOMIC WEIGHT
58.933

PHASE AT STP
Solid

Mining has always been a mixture of the magical and the mundane. From the earliest times it was dangerous and exhausting, often using slave labor. Miners risked tunnel collapses and entrapment, exposure to toxic and lung-clogging dust, and work-related injuries and deformities. Yet by burrowing into the underworld they were exploring a hidden realm where no one knew quite what the rules were, or what manner of beings awaited them.

In *De Re Metallica*, Georgius Agricola warns of one particular type of mineral—a kind of "pyrites," he says—that "has the distinctive property of being extremely corrosive, so that it consumes the hands and feet of the workmen, unless they are well protected." The German miners figured that this awful stuff was connected to subterranean beings, the gnomes and goblins that were thought to haunt mines and torment workers. They named it after the German word for such creatures: kobolds or *kobelt*, from which comes the element name cobalt. Like many metals, cobalt is only poisonous in large doses, and, in fact, our bodies need a small amount of it to stay healthy: it is a key element in vitamin B12. So, it's not clear whether it was actually cobalt ore that was so hazardous to the miners of Agricola's day; the danger might have come more from arsenic, minerals of which often accompany those of cobalt,

LEFT: Ulrika Pasch's portrait of Torbern Olaf Bergman, 1779, National Portrait Gallery, Mariefred, Södermanland.

OPPOSITE: The cobalt-based *bleu de Chartres* on the robes of the Virgin, stained glass window of Notre-Dame de la Belle Verrière, at Chartres Cathedral, twelfth to thirteenth centuries.

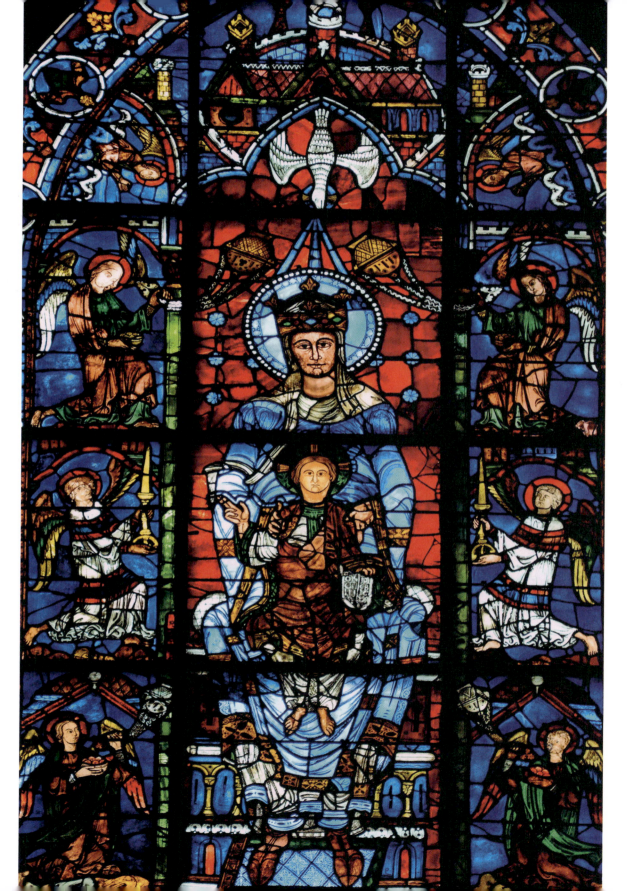

ABOVE: Subterranean creatures (kobolds) disturbing miners. Woodcut from Olaus Magnus's *Historia de Gentibus Septentrionalibus*, 1555, Chapter XXII, National Library of Norway.

nickel, zinc, and bismuth. The most notable feature of cobalt ores is that they can be a rich blue color, known in the Renaissance as *zaffre*—the word is related to "sapphire," although those blue gemstones get their hue a different way.

Zaffre was used to make the finest blue glass, simply by adding some of it to the glassmaking furnace. The Romans knew how to do that, but their method was lost in northern Europe during the Middle Ages—so the wonderful, blue stained glass in Gothic cathedrals such as Chartres was generally recycled cobalt-blue glass from Roman times. Indeed, there was a thriving trade in such relicts imported from the south, from Byzantium or the Islamic lands: a shipwreck in the Aegean Sea from the eleventh century carried a cargo of several tons of blue, green, and amber-colored glass shards, probably intended for sale to glassmakers in Europe. In the following century, at the dawn of the Gothic age, the German monk Theophilus

LEFT: Cobalt-blue glass ingot, fourteenth century BC, perhaps Syrian, from the Bronze Age Uluburun shipwreck, near Kas, Turkey, Institute of Nautical Archeology, Texas.

wrote of these rich blues that "there are also found various small vessels in the same colours, which are collected by the French…They even melt the blue in their furnaces…and they make from it blue glass sheets which are costly and very useful in windows."

A form of cobalt-blue glass was also used, finely ground, as a pigment by painters, often called smalt. It wasn't ideal: it had a gritty texture, and mixed with oils the blue was never quite as glorious as it was in a church window with the sunlight streaming through it. It wasn't until the nineteenth century that chemists discovered how to make better use of the rich blue of cobalt compounds: in 1802, the

Frenchman Louis-Jacques Thénard discovered how to make the compound cobalt aluminate, which was sold as the pigment cobalt blue.

Once again, we don't know when cobalt ores were first "reduced" to make the silvery metal of cobalt itself. But the first person to claim cobalt as an element was a Swedish chemist named Georg Brandt, who studied the blue cobalt ore and concluded in 1739 that it did indeed contain a previously unrecognized metal. Three years later he succeeded in isolating the element itself, which he found was magnetic—although pure samples of cobalt weren't made until 1780 by Brandt's compatriot Torbern Bergman. Brandt listed it alongside mercury, bismuth, zinc, antimony, and arsenic as a "half-metal"—further evidence that the roster of chemical elements was larger than had been suspected and raising the questions: why so many, and where did it end?

BELOW: The Glassmakers Guild making colored glassware. From *Surname-i Hümayun* (Book of the Imperial Circumcision Festival), 1582–1583, manuscript H. 1344, Topkapi Saray Museum, Istanbul.

ARSENIC

GROUP 15

33

As
Arsenic

Metalloid

ATOMIC NUMBER
33

ATOMIC WEIGHT
74.922

PHASE AT STP
Solid

Although arsenic is not itself a metal, it was closely associated with the new metals such as cobalt and zinc that were being mined during the Renaissance. Among the most common arsenic-containing minerals are orpiment and realgar, both of them forms of arsenic sulfide and both brightly colored. Orpiment is yellow, and was used as a pigment from at least the second millennium BC in ancient Egypt; the name comes from the Latin *aurum pigmentum* (pigment of gold), and this association with the most noble of metals led some painters to refer to it as King's Yellow. The Arabic word for arsenic, *al zarniqa*, meaning "gold-colored," is the root of the modern name of the element; in Greek, orpiment was called *arsenikon*.

Orpiment was a rare and expensive material, which only the most well funded of artists could use. And they knew from bitter experience that it was hazardous. In his craftsman's handbook of around 1390, the Italian artist Cennino Cennini warns that this pigment is "really poisonous," and advises that you should "beware of soiling your mouth with it." It was the same with realgar, which is orange. This was more or less the only pure orange pigment

BELOW: Orpiment and realgar, used for yellow and orange pigments, in the robes shown in Antoine Watteau's *The Italian Comedians*, ca. 1720, from the Samuel H. Kress Collection, National Gallery of Art, Washington, DC.

OPPOSITE: Parchment from Albertus Magnus's *De Mineralibus et Rebus Metallicis Libri Quinque*, in which he describes a recipe for arsenic, Italy, 1260–1290, manuscript 20, Iron Library, Schlatt.

Incipit prim⁹ liber mineraliu^m qⁱe de lapidib⁹. ca^r
demonstratione [...] p^m e de lapidib⁹ [...]
[...] a congelatione [...] Et in q^{uo} c^m [...]
[...] a congelatione et liq^{ue}fatione t^{er}tio z^m dif-
ferencias bi pullorib⁹ modis tor-
[...] methodo^rum visa e dem^m. p^{ri}a istii est
[...] qⁱa apparet apud res nr^as lapid^m ge douu
[...] et metallo^r^u et ea q^{ue} media st^{er} int^{er} eos ooo
sic^{ut} marcasida [...] aliud z quodlⁱ alia talia z ooo
[...] p^{ri} st^m no opposita p^{er} nos excisis ut po-
[...] a complexionata cristalla q^{ua}lita st^{er} id de his
[...] p^{er} st^{er}m methodo dd oct^{er}et. p^{ri}z est inde-
terminato uf inuentione simplicet eo^r^u.

De huia^{us} lib^{er}i a^ut^{er} no inuenim⁹ no excepto^r p^{re}ter
[...] qⁱ tradidit a nob^{is} de huia t. e. p^{er} lib. ar^{um} in q^uaⁿ
[...] de his no suffic^{er}. p^{ri}a qⁱ de lapidib⁹ z p^{er}ba
de metall⁹ z uisco de medija int^{er} ea facient⁹. et
[...] stione lapis q^{ua}pp^{er} p^{ri}a facilio^r c^{er} et mag⁹
manⁱfesta q^ua metallo^r. de lapid^m e naturalis p^hia
[...] qⁱ iph⁹ ponem⁹ z dem^m de
lapidib⁹ z st^{er} q^{ue} mag⁹ notata st^{er} disputabilia
[...] q^{uo} uisco^r lf deo^r
[...] in^{er}e qⁱ de lapi-
[...] qⁱrem⁹ in q^ue m^u lapid^m z p^{ri}u^m essia
[...] eo^r^u p^{ri}m⁹ z loca⁹ g^eratio^r z dem^m mod⁹
[...] inmixtionis lapid⁹. cu^m diuersitate coloru^m eo^r
[...] z uictut^{es} z ipis p^hic e durit⁹
[...] maioris minoris dolabilitas et dolabilitas po-
[...] z structio g^enit⁹ z fluentur z^c. si bi i q^{ue}g^{bz}
lapides no solu^m i sit^u mo. st^{er} z signa iudeu^r
[...] hic diuersitat⁹ st^{er} a q^{ua}da mt^{er} auctoru^m
[...] iph⁹a in re gⁱ no deo^rib⁹. st^{er} do^{ct}ozⁱ lapidi gⁱlib⁹ to-
[...] tract⁹ summat^{es} se d^mt de lapidib⁹ fe^cit ali
[...] Gl^{er}so⁹ hui⁹ eui⁹ce⁹ rex arabi⁹ q^ua^uso
[...] de auctore Ioseph de lapidib⁹ tr^et⁹ p^{ri}a
[...] de uno lapid^e t^ractaut. nim⁹ infide a-
[...] z nocisa t^radidit p^hia hystoria fid^u no sapi-
[...] libⁱ lapid⁹ qⁱ assig^{ar}ut⁹ uoc^{er} ozⁱ uoc^{er} oi^m ho^r
[...] dim⁹ eo qⁱ sta^{er} e rei uisa⁹ de octa qⁱ iph^m
[...] certorib⁹ nos colligⁱo opteat. si^{er}
[...] e fo^{er} sciens rei⁹ lapid⁹ z pleio^res qⁱ ipis
[...] st^{er}t^{er} nt⁹ ut essia⁹ p^{er}m z^{er}archa ip⁹ eo^r
[...] p^{er} t^ertia i qⁱto methodo^r z qⁱst^{er}de in no⁹ ii^di
[...] osten^{er}de qⁱr aliq^u isto^r tm mⁱnut⁹ tc ^tali

[column 2]

an q^{ua}le p^{ri}ud⁹tot^u mediane e⁹ ch⁹ip⁹r uocat altic-
[...] me^a eui⁹ e gⁱaliom⁹. eo^r z de lta eo^r mⁱfesto^r
[...] eo^r mⁱfesta eo^r detegat⁹. s^{er} p^{ri}o inuenibo^r eo^r
excisis z q^{ua}lⁱ uisp^{er}z eo^r z ip⁹i st^{er} ost^{er}nut⁹. sg^m qⁱ st^{er}
[...] inquirere dz^m lapid⁹ z sp^{er} sia a^{li}e z^{er}or
[...] be z aqⁱb⁹ de qⁱb⁹ iq^{ui}nt⁹. at^{er}ma lapid^{er} uo
[...] dc^{er}. do idq⁹ eui⁹p^{er}t signe z^{er}b uocat⁹. eo^r z sia^m
[...] dia qⁱ eui⁹p^{er}t signe sic^{er} sulph⁹ z argedu^m uiu^m
[...] ex qⁱb⁹ diu^{er}so^r color⁹ si^{er} ea qⁱ uocat⁹ lapidos uo
[...] cu^m sp^{er}m z^{er}um z acus. Alⁱb⁹ ei sia⁹ i q^{ui}rere debu⁹ q^{ue}
[...] das ualde silat⁹ t^{er}rib⁹ z silat⁹ modo qⁱ i signal⁹ hu⁹
[...] n^{us} ili tenebim⁹ z ide do plus s^{er}if⁹ z p^{ri}er t^{er}tiu^m et
[...] mixta caplⁱ tot⁹ op^{er}ai eu^{er}. et ^mista deptualib⁹ ly a^m
[...] at t^{er}tat⁹. ozⁱ nor p^{er} excisig^{er} z essⁱlb⁹ cognoisc^{er}d⁹ nt⁹
[...] isto^r z^{er}uilis deue t^{er}ire t^{er}al⁹ eo^r z p^{ri}ones eo qⁱ fig-
[...] z essⁱs nob⁹ st^{er} mⁱfesta. z in^{ui}sid aut^{er} a^d nt⁹ de qⁱb⁹ i
[...] oib⁹ istar⁹ lib. fe^cim⁹ ultim⁹. c^{er}ut⁹ p^{ri}cede⁹ eo^s a^{cts} i
[...] dea a^d essⁱlb⁹ z uictut^{es} z signa eo^r qⁱ i t^{ali}b⁹ d⁹ a ost^{er}
[...] sis⁹e mag⁹ z qⁱ a^d no⁹ mⁱfesta si^{er} id p^{er} p^hia eo^r z
[...] dc^{er}t⁹. Ier o^rdine ip⁹i⁹ lib ad sc^{er}ipto⁹ deia a^d lib⁹ a^t
[...] dt^{ui}one z fine nrⁱ lib metho^r v^o dix^{er} do qⁱb⁹ p^{er}
[...] de qⁱb⁹ p^{ri}s^{er} et do⁹. c^{er} ei lapid⁹ z metallo^r gⁱa st^{er}
[...] omⁱnia pl⁹i p^{er}late⁹ fuci e⁹. diu^{er}sist⁹ pauc⁹ z^{er} fe-
[...] medie z solut⁹ flos z^{er} t^ruct⁹ omⁱnia. au^{er} st^{er} a^u p^{ri}mⁱ
[...] omⁱmonia ip⁹ e⁹ t^{er}tans de lapidib⁹ z e^{er}te mⁱn^a
[...] lib⁹ q^{ue} deco^r9 aut^{er}s

Incipit⁹ de g^e^ro^r t^{er}tare de lapid⁹ nrⁱ t^rde duu⁹ o^mi⁹
lapidis mⁱn^a e⁹ a^u sp^{er}m e^dl t^{er}re a^u sp^{er}m qⁱ d⁹
aqⁱ. uis^{er} et i^{er} lapidib⁹ alt^{er}u^m isto^r elo^r. z^{er} t hui⁹ a
[...] qⁱb⁹ q^{ua}d st^{er} a^d diu^{er}t int^{er}. c a^d t^{er}re sf⁹ diu^{er}su⁹
[...] a signa e fe^{er}e oi^m lapid⁹ scⁱu mⁱg^r sb^aqⁱ qⁱ hi^ed^u
[...] re t^{er}tio m^o duu⁹ c^{er} st^ad⁹ eo^r z mⁱ. si. m^{er} c^{er}t^{er} d^ma uai⁹
[...] eo^r st^{ro}a⁹ c^{ta}la paldi⁹no naturet⁹ st^{er} a^ds⁹. inu^{er} a^u
[...] lit lapid⁹ q^{ui} natui^t ni sit st^{ro}ut ut⁹ a^dust⁹ z p^{er}a^d
[...] ustione sp⁹o^{sol} sn⁹ si^{er} pianez^{er} flap⁹ qⁱ euo^mit⁹
[...] dine ut⁹ igis uiulet⁹ q^uiⁿ f^{er} ts^{er} p^{ri}stiue redig^atur⁹
[...] p^{ri}us sp⁹ sb^aqⁱ mⁱg^rt⁹. Ad⁹u⁹ d^{er} t sit i^{er} lapidib⁹ p^{er}pi⁹
[...] eui⁹a aqⁱ. nl⁹ eget brea⁹. aut^{er} histd⁹ui ad^miuet⁹. ni in
[...] set⁹ sb^aqⁱ c^{er}till⁹ z b^{er}rill⁹ si^{er} u^o mⁱg⁹ placer⁹
[...] z^{er}. z qⁱb⁹ p^{er}l uil⁹ As⁹ histd⁹ui⁹ e a^d a^dl⁹ z^{er} lapides op⁹
[...] zglu^{er}at⁹ z reuil⁹b⁹ z uesto⁹ a^{li}ut⁹ gⁱlam⁹ ex uisdos⁹i⁹
[...] grossa⁹ z dc^{er}t⁹ lut^{er}e. ozⁱ p^{er}t qⁱ ut⁹ sit mⁱ lapid⁹. sc^{er}d
[...] hie z do^{ct}or⁹ de hui⁹ lapidib⁹ z qⁱb⁹ c^{er} sca^m in⁹i⁹i
[...] ale p^{ri}e p^{er} qⁱ il⁹ histd⁹ no solu^m t^{er}e e⁹ i^{er} macri⁹ ni ati

OPPOSITE: The week before this engraving by John Leech appeared in an issue of *Punch*, the chemist A. W. Hoffman published an article stating that green dresses, wreaths, and artificial flowers made with the arsenic-containing compounds copper arsenite and copper acetoarsenite (Scheele's green and Paris/emerald green) were toxic. Wood engraving, *The Arsenic Waltz, The New Dance of Death (Dedicated to the Green Wreath and Dress-Mongers)*, *Punch*, February 8, 1862, Wellcome Collection, London.

LEFT: The pigment known as Paris or emerald green contained arsenic and was used in wallpaper as well as by artists. This is William Morris's first *Trellis* wallpaper (designed in 1862 and then produced in 1864), The Metropolitan Museum of Art, New York.

artists had until the nineteenth century—otherwise they had to mix the color from red and yellow—and so the temptation to use it was strong. But Cennino says "there is no keeping company with it."

Agricola mentions that German miners working veins that contained cobalt, zinc, and silver might often encounter a *cadmia metallica*—a rather vague term for an unknown ore—that smelled of garlic, and which we can therefore conclude contained arsenic compounds.

Like antimony, arsenic is, in fact, a metalloid: it is silvery gray, but conducts electricity poorly. Again, we don't know when pure arsenic was first separated from its natural ores. The Greek writer Zosimos of Panopolis in the third century BC, who we encountered earlier as one of the fathers of

Western alchemy, described how *sandarach*—an old name for realgar—could be heated to make (as we would now recognize it) arsenic trioxide, from which the oxygen could be removed by heating with oil, to make arsenic. Yet it is always very difficult to know exactly what was going on in these old alchemical accounts—and the experiments here are pretty dangerous. An account of what sounds like the preparation of pure arsenic by this route (or something similar) appears in a manuscript that is attributed to the thirteenth-century German Dominican friar and experimenter Albertus Magnus.

Arsenic has a reputation as the poisoner's favorite—or it was, until a chemical method called the Marsh test was developed in the 1830s to detect

THE ARSENIC WALTZ.
THE NEW DANCE OF DEATH. (DEDICATED TO THE GREEN WREATH AND DRESS-MONGERS.)

traces of it in a post-mortem. This led to some high-profile exposures of arsenic poisoning in one of the first demonstrations of forensic science. The trial of Mary Ann Cotton, who was convicted in 1873 of murdering her stepson Charles Edward Cotton by arsenic poisoning, caused a public sensation. It seems that she also killed three of her former four husbands in this way to collect their life insurance; the idea of a female serial killer sweetly putting arsenic-laced sugar into her husbands' tea captivated the public imagination. Cotton was exposed by the Marsh test, which revealed traces of arsenic in her stepson's autopsy.

Even though the toxicity of arsenic was well known by then, two arsenic-containing copper compounds were widely used as green pigments throughout the nineteenth century. One, Paris green or emerald green, was used not only in artists'

paints but also to print patterned wallpaper, leading to concerns in the 1860s that arsenic fumes emitted from damp rooms papered in the stuff were killing people, including children, in their sleep. One of the main sources of arsenic for these green pigments in the later part of the century were the mines in Cornwall owned by William Morris, whose floral wallpaper designs were much in demand. Despite his reputation as one of the leaders of the Arts and Crafts movement, which advocated a return to traditional methods of manufacture in the face of industrialization, Morris profited from the production of this lethal compound. Furthermore, legend has it that the green-painted walls of Napoleon Bonaparte's lodgings during his exile on St. Helena hastened his death.

MANGANESE

GROUP 7

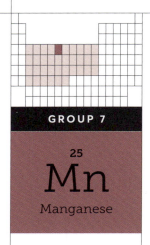

25

Mn

Manganese

Transition metal

ATOMIC NUMBER
25

ATOMIC WEIGHT
54.939

PHASE AT STP
Solid

There are metals whose properties humans have exploited since long before they were recognized as elements. Cobalt is one: the source of the glorious blue glass of the era of Gothic cathedrals. Another element that proved invaluable to glassmakers of ancient and medieval times was manganese.

It's a curious fact that making colored glass was always easier than making it clear. Glass is basically a form of silica—the same, in terms of its elements, as the clear mineral quartz, made from silicon and oxygen, but with the atoms joined together in a more disorderly fashion. It was made from around 2500 BC by melting sand with ashes or with natural soda (a mineral the Romans called *natron*); adding small amounts of other metal ores could impart a certain color. But even without those additives, ancient glass tended to be lightly colored because of the impurities present in the sand: trace amounts of other elements, some of which form compounds that are strongly colored. Iron was one of these, making glass a pale green, yellow, or red. Another was the element manganese, mineral ores of which are relatively common. Depending on how well aerated the glassmaking furnace was, the presence of manganese could turn glass purple or yellow—or, with some iron present too, a gorgeous reddish saffron. Glassmakers didn't understand what caused these colors, and it wasn't easy to control them, but such richly tinted glass was highly prized.

RIGHT: Laboratory equipment. Title page from Carl Wilhelm Scheele's *Chemische Abhandlung von der Luft und dem Feuer*, Upsala: Magn. Swederus, 1777, Smithsonian Libraries, Washington, DC.

On the other hand, these artisans also discovered that manganese ore can strip glass of its color and leave it as transparent as quartz—a product that, in the first century AD, the Roman writer Pliny the Elder says was the most highly prized. Glassmakers might add to their kilns a small amount of a mineral that came to be known as pyrolusite, from the Greek meaning "fire-washer"—for in the fiery furnace it would cleanse the glass of all tint. In the Middle Ages, this substance was often called "glass soap." It's an unexpected result, since pyrolusite itself is black, and in fact was used as a black pigment by some cave artists at least 17,000 years ago. In chemical terms the mineral is manganese dioxide—and the metal will scrub the glass of color.

The Flemish chemist Jan Baptista van Helmont wrote about this property in a book of 1662, saying that the substance will "draw anything out of glass thoroughly boiled or melted by fire; for a very small fragment thereof, being cast into a mass or good quantity of glass while it is in boiling, of green or yellow makes it white." He refers to the manganese mineral as a "lodestone," meaning that it is a magnet—for indeed it is, slightly. For this reason, it was referred to in the Middle Ages as *magnesia*, a general term for a magnetic stone. In his mid-sixteenth-century treatise on metalworking, an Italian named Vannoccio Biringuccio transposed some of the letters in this name, calling the mineral *manganese*. It was by this name that pyrolusite became known for the next two hundred years or so.

For chemists of the late eighteenth century, the question was: what was actually *in* minerals like this? No longer content to think of them vaguely as a kind of "earth," they wanted to figure out which elements they contained. The great Swedish chemist and apothecary Carl Wilhelm Scheele tried to find out in the case of pyrolusite, which he suspected of harboring a new element—but he couldn't isolate one. That discovery fell to Johan Gottlieb Gahn, an assistant of the eminent Norwegian chemist Torbern Bergman, in 1774. (It is quite possible, however, that pure manganese was first produced three years earlier by a young Viennese chemist, Ignatius Kaim.) The glassmakers' magnesia, wrote Bergman, is actually the "calx of a new metal," meaning the

ABOVE: A "Runge pattern" made using chromatography by the German chemist Friedlieb Ferdinand Runge for his 1858 book *Der Bildungstrieb Der Stoffe*. The colors are created by various metal salts, including manganese oxides. Oranienberg, Germany, 1858, Plate 16, Science History Institute, Philadelphia.

substance formed when that metal is burned in air. He reported how Gahn had managed to obtain a "regulus" from this magnesia, meaning a blob of the pure metal itself. However, Bergman then went on to sow the confusion that has bedeviled student chemists ever since—because he calls this metal not manganese (which in truth it was) but the perplexingly similar *magnesium*—another element entirely, and yet to be discovered. Here again is an indication of how hard it was, as the elements proliferated, to tell them apart, especially when so many of them seemed to be silvery metals. But there were many more of those still to come.

TUNGSTEN, PLATINUM, AND PALLADIUM

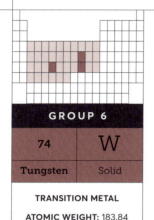

GROUP 6

74	W
Tungsten	Solid

TRANSITION METAL

ATOMIC WEIGHT: 183.84

GROUP 10

78	Pt
Platinum	Solid

TRANSITION METAL

ATOMIC WEIGHT: 195.08

GROUP 10

46	Pd
Palladium	Solid

TRANSITION METAL

ATOMIC WEIGHT: 106.42

Mining and mineralogy drove much of the element-discovery from the sixteenth to the mid-eighteenth centuries. It wasn't so hard to spot a new mineral—say, from its color, its density, the shape of its crystals. And it had long been known that minerals were a source of metals. On the other hand, there was ample scope for confusion: different minerals could contain the same metal, a single mineral could contain more than one metallic element, and one type of "stone" might be confused for another. The nomenclature was a nightmare: minerals had different names in different places, there was confusion of the mineral with the metal, and alchemical terms and ideas—not least about the transmutation of metals—still lingered.

In Germany there was a substance often found in conjunction with tin ore that, in the mid-sixteenth century, miners called Wolfrumb or Wolffram— or perhaps *Wolffschaum* (wolf foam) or *Wolffshar* (wolf's hair), as it was black and formed fibrous crystals. It was said to make tin brittle when it was present in the ore. In 1747, the German mineralogist Johann Friedrich Henckel called this troublesome stuff *lupus Jovis*: "Jupiter's wolf," Jupiter being the planet that alchemists used to associate with tin. (It's not clear whether he meant to imply that this "wolfram," as it became known, devours tin like a wolf—but that's what some later writers fancifully assumed.)

Around the middle of the eighteenth century, Swedish mineralogist Axel Fredrik Cronstedt reported a "heavy stone"—in Swedish, tungsten. His compatriot Carl Wilhelm Scheele studied it, and may even have isolated from it the metal now known by that name. But the discovery of the metal tungsten is usually attributed to the Spanish chemists Juan José and Fausto d'Elhuyar, who looked at both the mineral then known as tungsten and also at wolfram—Juan José having recently visited Scheele. In 1785, the English-language version of their report on "a chemical analysis of wolfram," and the new metal it contained, made the discovery widely known. It was clear that the same metal was in both minerals, and the English text named this "wolfram" too. (Today the mineral ore is called wolframite.) But in France the metal was called *tungstène*. When the Swedish chemist Jöns Jacob Berzelius began to assign elements one- or two-letter symbols in the early nineteenth century, he decided that it should come from "wolfram": W. But by and by, the name tungsten stuck in England— and that is why tungsten is so puzzlingly denoted in the periodic table today.

Tungsten is a very dense metal—hence the Swedish name of the ore from which the word derives. Density was one of the few dependable clues scientists had that they had found a new metal. In the case of the dense, silvery metal platinum, that discovery was remarkably easy—for platinum is one of those rare metals that can occur in nature in its "native" elemental form. It was first identified in the early eighteenth century in alluvial (riverborne) deposits of silver in South America, especially around the Pinto river in Colombia,

LEFT: The extraction of metals from minerals. Woodcut from the title page of Lazarus Ercker's *Beschreibung Allerfürnemisten Mineralischen Ertzt unnd Bergkwercks Arten*, Frankfurt: Joannem Schmidt, 1580, Science History Institute, Philadelphia.

which explains its name: from the Spanish *platina*, meaning "little silver." Like silver and gold, platinum is a rather unreactive metal and so does not tarnish easily. This made it ideal for jewelry—in fact, the native South Americans around Colombia and Ecuador had worked it for hundreds of years for decorative purposes before the Spanish arrived.

This natural platinum was impure, however: it was really an alloy containing iron and tiny amounts of other metals. It's not clear who first realized that platinum is an element in its own right, although it became more widely known in Europe when a Spanish administrator, explorer, and scientist named Antonio de Ulloa was captured by the English on a French ship sailing back to Spain, and shared what he knew with the London scientific establishment the Royal Society. (They eventually made him a Fellow.) Other Europeans began to investigate platinum in the 1750s, and it took its place in the roster of elements.

Among those scientists who studied platinum were the London-based chemists William Hyde Wollaston and Smithson Tennant, who collaborated closely. Platinum is hard to melt—its melting point of 3,222°F is one of the highest of all metals—but it can be dissolved in a mixture of hydrochloric and nitric acids (known as *aqua regia*, "water of kings").

Wollaston found a way of precipitating platinum from this solution, and discovered that some other substance was left behind in the liquid that could be precipitated separately as a yellow solid. When he heated it, he found it decomposed to give a silver metal. Mindful of the old association of metals with the heavenly bodies, he named this one after the asteroid Pallas, which had just been discovered by astronomers in 1802: it became palladium.

Wollaston did not announce his discovery at once, but sold this "new silver" through a London mineral merchant, at six times the price of gold. He finally revealed the new metal in a paper presented to the Royal Society in 1805, in which he excused his secrecy by claiming his right to "take advantage" of the metal (although there were few takers for it). Meanwhile, Wollaston's partner Tennant studied the black residue left behind when platinum dissolves in *aqua regia* and found it contained two further new metals: osmium, which is even denser than tungsten, and rhodium. Along with iridium and palladium, these are known as the platinum-group metals.

OVERLEAF: Mining ore. Depicted on a panel of Hans Hesse's *Annaberg Mountain Altar*, St. Ann's Church, Annaberg-Buchholz, Germany, 1522.

URANIUM

GROUP N/A

92

U

Uranium

Actinide

ATOMIC NUMBER
92

ATOMIC WEIGHT
238.03

PHASE AT STP
Solid

William Wollaston's naming of palladium shows how, long after they had abandoned the alchemical and somewhat mystical belief in a "correspondence" between the heavenly bodies and the metals, chemists retained an affection for the idea. It also occurred to the German mineralogist Martin Klaproth when, in 1789, he named another dense metallic element.

Klaproth was studying the black mineral pitchblende, which was commonly found in the silver mines of Bohemia. Its name derives from its black color, like tar or pitch, and the German word *blende*, literally meaning "to deceive," which was given to any stone that seemed heavy enough to contain a metal but from which none could be extracted.

Klaproth didn't actually extract a pure metal from pitchblende. He dissolved the mineral in nitric acid, and then found that adding an alkali to the solution precipitated a yellow substance. On heating it, he made a black powder, which he assumed was some new metal and gave it a name. "As the discovery of new planets has not kept pace with that of metals," he wrote, "the metals newly found out have been deprived of the honour of receiving their names from planets, like the older ores." However, a new opportunity to bestow that honor had just arisen, for in 1781 William Herschel in England had discovered a new planet using his telescope and named it after the Greek sky god Uranus. So, for his new element, Klaproth announced, "I have chosen [the name] *uranite* (*Uranium*), as a kind of memorial, that the chemical discovery of this new metal happened in the period of the astronomical discovery of the new planet *Uranus*."

The road to radioactivity

Klaproth's black powder was not really uranium, however, but uranium oxide. The pure metal was not isolated until 1841 by the French chemist Eugène Peligot in Paris. One of the first uses of the metal was as a coloring agent for glass and pottery. Glass containing uranium oxide was tinted a fluorescent yellow-green, and later a uranium compound was used for a rich orange pottery glaze, which was popularized in the 1930s and 1940s by the Fiesta range of crockery. In fact, a scientist working at Oxford in 1912 claimed that some of the pale green pieces in a glass mosaic from an Imperial Roman villa near Naples contained small amounts of uranium that must have been deliberately added (as the ore). It's still debated whether that finding is genuine, as no other Roman glasses containing uranium are known—although some scientists have suggested the uranium may have come from Cornish mines in Roman Britain.

OPPOSITE: Silver mining in Kutná Hora, Bohemia (now the Czech Republic). From an illuminated choirbook, 1490, Sotheby's London.

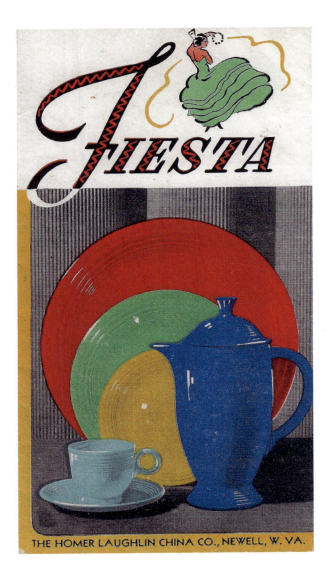

ABOVE: Advertising brochure for uranium-glazed Fiesta-ware, Homer Laughlin China Company, 1937. In 1943, the red glaze was discontinued as uranium was needed for the war effort.

For many decades, then, uranium was a curiosity. Its ore pitchblende proved to have the unusual property that, if illuminated with light and then placed in the dark, it would glow: behavior called

phosphorescence. This was seen as a novelty, good for parlor-room tricks but little else. But it interested the French physicist Alexandre-Edmond Becquerel, who studied the phenomenon carefully in the mid-nineteenth century.

In 1895 the German scientist Wilhelm Röntgen reported a new kind of "emanation" that he called X-rays, which would leave an imprint on photographic emulsion, like light—but which could pass right through solid objects, such as flesh. It was soon found that these X-rays could induce phosphorescence in some substances, too. In early 1896, Becquerel's son Henri wondered if phosphorescent materials like uranium salts might actually *emit* X-rays. He wrapped some photographic plates in black paper to protect them from exposure to light, then placed various phosphorescent materials on top of them and exposed the materials to sunlight to excite their phosphorescence. None left any imprint on the plates—except the uranium compounds.

At first, Henri Becquerel assumed that the effect was due to the light-induced phosphorescence of the uranium salts. But then he happened to leave some of the plates in a closed drawer for a few days, the weather being too gloomy in February to produce much sunlight. Some instinct prompted him to develop the plates anyway, even though they hadn't been exposed to uranium phosphorescence. To his surprise, Becquerel found that they still produced a photographic imprint. He concluded that the uranium compounds were themselves emitting a different kind of ray, which were dubbed uranic rays.

Becquerel had discovered that uranium is radioactive—though it was left to others to understand what that really meant. And that's another story of elemental discovery.

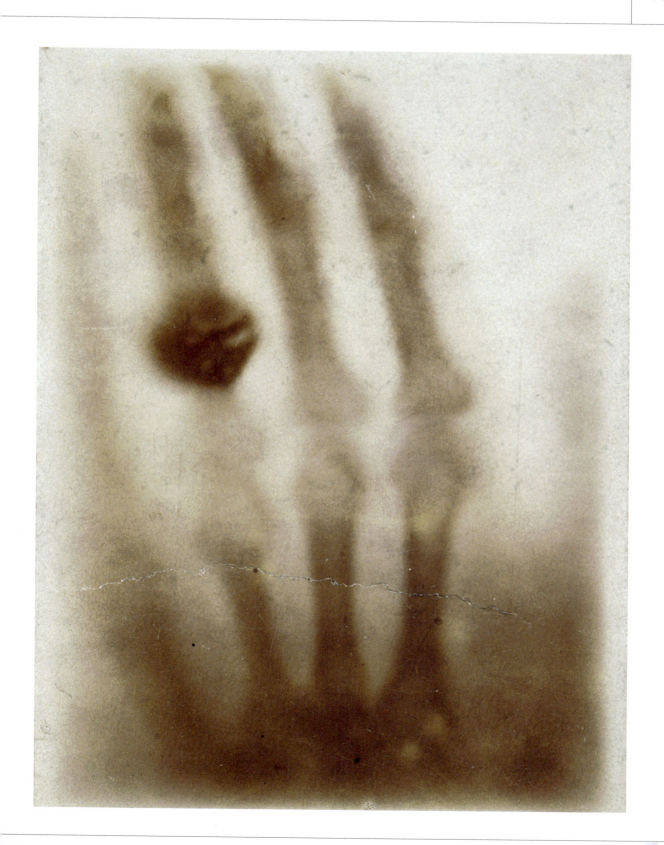

A Synopsis

OF THE

CHEMICAL CHARACTE

Adapted to the NEW Nomenclature,

By Mess.rs HASSENFRATZ and ADET,

Systematically arranged by W. Jackson, Practical Ch

The CHEMICAL CHARACTERS of the NEW NOMENCLATURE are divided into two Clases simple & compound, cont
six Genera & fifty five Species, each Genus has a Sign proper to itself which with some Modifications exprfses the different Species in ea
and by the Combinations & Positions of the six Generical Characters the constituent principles & proportions are exprefsed of all compound

GENERICAL CHARACTERS

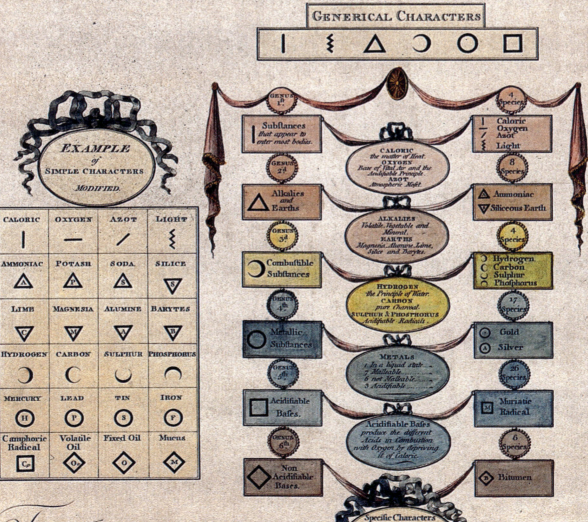

EXAMPLE
of
SIMPLE CHARACTERS
MODIFIED.

CALORIC	OXYGEN	AZOT	LIGHT
AMMONIAC	POTASH	SODA	SILICE
LIME	MAGNESIA	ALUMINE	BARYTES
HYDROGEN	CARBON	SULPHUR	PHOSPHORUS
MERCURY	LEAD	TIN	IRON
Camphoric Radical	Volatile Oil	Fixed Oil	Mucus

GENUS 1.st Substances *that appear to enter most bodies.*

GENUS 2.d Alkalies *and* Earths

GENUS 3.d Combustible Substances

GENUS 4.th Metallic Substances

GENUS 5.th Acidifiable Bases.

GENUS 6.th Non Acidifiable Bases.

CALORIC *the matter of Heat.* OXYGEN *Base of Vital Air and the Acidifiable Principle.* AZOT *Atmospheric Moist.*

ALKALIES *Volatile, Vegetable and Mineral.* EARTHS *Magnesia, Alumine, Lime, Silice, and Barytes.*

HYDROGEN *the Principle of Water.* CARBON *pure Charcoal.* SULPHUR & PHOSPHORUS *Acidifiable Radicals.*

METALS *1 In a liquid state. 7 Malleable. 6 not Malleable. 3 Acidifiable.*

Acidifiable Bases *produce the different Acids in Combustion with Oxygen by depriving it of Caloric.*

4 Species Caloric Oxygen Azot Light

8 Species Ammoniac Siliceous Earth

4 Species Hydrogen, Carbon, Sulphur, Phosphorus

17 Species Gold Silver

26 Species Muriatic Radical

6 Species Bitumen

Specific Characters *for Substances that may be discovered*

To J. . . . . M.D. F.R.S.S. . . . This Table of Chem

CHAPTER FIVE

CHEMISTRY'S GOLDEN AGE

LEFT: *A Synopsis of the Chemical Characters.* Engraved broadsheet by H. Ashby after W. Jackson, for Hassenfratz & Adet, 1799, Wellcome Collection, London.

CHEMISTRY'S GOLDEN AGE

The eighteenth century began with chemistry still recognizably linked to the alchemical tradition, not least in the central role given to phlogiston as the "principle of combustion." By the start of the nineteenth century, chemistry looks recognizably like its modern self: many of the familiar elements are there, the idea of atoms and molecules was just emerging, and so was a dim awareness of the notion of a chemical bond that glues atoms into those molecular unions.

The eighteenth century was the age of what is sometimes called the Chemical Revolution, a slightly delayed counterpart to the seventeenth-century Scientific Revolution that gave us the basic outline of modern science as a body of knowledge describing the world in non-magical terms, governed by mechanics, expressed in mathematical laws, and deduced by careful and systematic experiment. Both terms offer too simplistic a picture, to be sure—it's not clear that "revolutions" are ever the right way to think about the advance of science, in which old, mistaken ideas always rub shoulders with bold new theories and two steps forward are typically accompanied by one step back. Human thought doesn't ever evolve without mess and argument, error and confusion.

Still, this period was undoubtedly transformative for chemistry. What is sometimes neglected in that development, though, is the practical side of the science. The Industrial Revolution was gathering pace, creating unprecedented demands for new substances and methods: techniques of metal extraction from ore; dyes, bleaches, and mordants (fixative agents) for the textiles industry; and paints, paper, inks, soaps, perfumes—all the materials needed to make life more agreeable to the emerging middle classes. The relationship of science to technology does not simply involve the conversion of new ideas to lucrative opportunities; technologies throw up challenges that lead to fresh thinking and discovery.

It's mostly in experimental investigations that science and technology meet—and chemistry has always been the paradigmatic experimental science. In the eighteenth century, chemists began to get to grips with how the elements combine, and how they can be persuaded to adopt new configurations. Chemistry became quantitative: it didn't just matter what would react with what, but in what proportions. The chemists' apparatus had to include not just the furnace and the retorts and other vessels in which they cooked and distilled the elements, but the balance for carefully weighing quantities before and after a reaction. That attention to detail was essential for getting a better understanding of chemical processes, but it was also important for industries that demanded efficiency and didn't want to see precious materials go to waste.

This quantitative turn of thought was reflected in the way chemists started to think about how elements combine. Old notions of elements having a kind of "love" for one another, so that they might join in chemical marriage, gave way to the concept of "affinity," or what the French chemists called *rapport*. Chemists drew up increasingly elaborate affinity tables showing how readily different

1. Acid Spirits	2. Marine Acid	3. Nitrous Acid	4. Vitriolic Acid	5. Absorbent Earth	6. Fixed Alkali	7. Volatile Alkali	8. Metallic Substances	9. Sulphur	10. Mercury	11. Lead	12. Copper	13. Silver	14. Iron	Regulus of Antimony	Water
Fixed Alkali	Tin	Iron	Phlogiston	Vitriolic Acid	Vitriolic Acid	Vitriolic Acid	Marine Acid	Fixed Alkali	Gold	Silver	Mercury	Lead	Regulus of Antimony	Iron	Spirit of Wine
Volatile Alkali	Regulus of Antimony	Copper	Fixed Alkali	Nitrous Acid	Nitrous Acid	Nitrous Acid	Vitriolic Acid	Iron	Silver	Copper	Lapis Calaminaris	Copper	Silver Copper Lead	Silver Copper Lead	Neutral Salts
Absorbent Earths	Copper	Lead	Volatile Alkali	Marine Acid	Marine Acid	Marine Acid	Nitrous Acid	Copper	Lead						
Metallic Substances	Silver	Mercury	Absorbent Earths		Acetous Acid		Acetous Acid	Lead	Copper						
	Mercury	Silver	Iron		Sulphur			Silver	Zinc						
			Copper					Regulus of Antimony	Regulus of Antimony						
			Silver					Mercury							
	Gold							Gold							

A TABLE of AFFINITIES BETWEEN SEVERAL SUBSTANCES, BY MR. GEOFFROY.

elements are apt to combine. One element might displace another if its affinity for the resulting combination was greater. There were laws governing the behavior of the elements.

Nowhere was the drive toward quantification greater than in France, where from the 1760s the chemist Antoine Lavoisier made very careful measurements of the proportions of elemental combination and separation. Lavoisier was one of the main proponents of the principle of chemical analysis (meaning literally "splitting apart"): he would deduce which elements a compound contained by breaking it down. In this way, Lavoisier furnished chemistry with a clear way to define elements: they are, he said, substances that cannot be split into simpler ones by chemical reactions.

In 1787 Lavoisier, with his colleagues Louis-Bernard Guyton de Morveau, Claude-Louis Berthollet, and Antoine Fourcroy, published *Méthode de Nomenclature Chimique*, a textbook that announced their new vision of chemistry by proposing to replace old, alchemical names for substances with new ones. "Oil of vitriol" became sulfuric acid, and "flowers of zinc" became zinc oxide. And, crucially, chemical names would be derived from the elements of which the substance in question was composed. Lavoisier and his followers literally rewrote chemistry.

ABOVE: "A Table of Affinities between Several Substances." From Pierre Joseph Macquer's *A Dictionary of Chemistry*, Vol. I, London, England: Peter Elmsley, 1777, Science History Institute, Philadelphia.

Two years later, Lavoisier assured the precedence of his new system with his great book *Traité Élémentaire de Chimie* (An Elementary Treatise on Chemistry). Here he compiled a new list of elements that contained no fewer than thirty-three of them, bearing the names Lavoisier insisted on. The book also described the practical art of chemistry, making it the standard textbook of chemical instruction for decades to follow—and implanting Lavoisier's vision of chemistry from the moment new students began to learn the topic.

Whatever accolades his accomplishments rightly drew upon him, Lavoisier did not enjoy them for long. As well as a scientist, he was also a tax collector and an administrator in the Gunpowder Administration of Louis XVI of France. So when the French Revolution exploded in 1789, he was a prime target for the revolutionary witch-hunters during Robespierre's Reign of Terror. In 1793 he was accused of being a traitor to the First Republic, and in May 1794 he was sent to the guillotine. Having instigated one revolution, he fell foul of another.

HYDROGEN

GROUP 1

1

H

Hydrogen

Non-metal

ATOMIC NUMBER
1

ATOMIC WEIGHT
1.008

PHASE AT STP
Gas

Deciding who first discovered an element can be tricky. Is it the first person who saw some sign of it, or who first separated and isolated the pure element? Or the first who knew that what they had made was indeed an element, and not just a new variety of another substance?

We are confronted with this question when it comes to hydrogen. Its discovery is usually attributed to the English scientist Henry Cavendish, who made and described it in 1766. But Cavendish himself believed that what he had found was a particular type of "air"—so not exactly an element at all.

Cavendish wasn't the first to make hydrogen in any event. Cosmically speaking, hydrogen is everywhere. It's far and away the most abundant element in the universe: fully nine out of every ten atoms in the cosmos is a hydrogen atom, and most stars are made predominantly of this element. Since it's the main ingredient of the clouds of gas from which stars and planets form, it also dominates the atmospheres of the gas-giant planets such as Saturn and Jupiter, making up 80–96 percent of their gaseous blankets. In contrast, there's hardly any pure hydrogen in the Earth's atmosphere, because it is the lightest of all elements and the planet's gravity can't hold onto it. It makes up around 13 percent of the atoms in the Earth's crust and surface, but almost all of that is bound up in chemical compounds, especially water.

The most common way to make pure hydrogen is to pull the atoms from water molecules. This is more or less what happens if you dissolve a fairly reactive metal in an acid. Robert Boyle carried out that reaction in 1671 using hydrochloric acid and iron filings—but others almost certainly did it before him. The hydrogen gas bubbles off—it's lighter than air—and can be collected. Boyle discovered that this gas is very inflammable, igniting with a pop and creating a bright light. To him and his contemporaries, it was "inflammable air." So why award the discovery to Cavendish almost a century later? Well, Cavendish was a supremely careful experimenter, who made accurate measurements of his studies with gases and looked closely at their chemical behavior: he took this "air" seriously as a substance in its own right.

Cavendish was also—even allowing for science's reputation for accommodating eccentricity—a deeply odd man. The millionaire grandson of a duke, Cavendish, like many wealthy "gentleman philosophers" of his day, carried out his studies privately from a home laboratory, which he built in Clapham in south London. Having no time for company or custom, dressing shabbily and never marrying, he could be found shuffling from room to room of the Royal Society, London's premier scientific institution, shunning discourse and reportedly emitting a "shrill cry" from time to time. One of his contemporaries called Cavendish "shy and bashful to a degree bordering on disease."

Cavendish made hydrogen in much the same way as Boyle, by reacting iron filings with acid (sulfuric). He interpreted its inflammability according to the

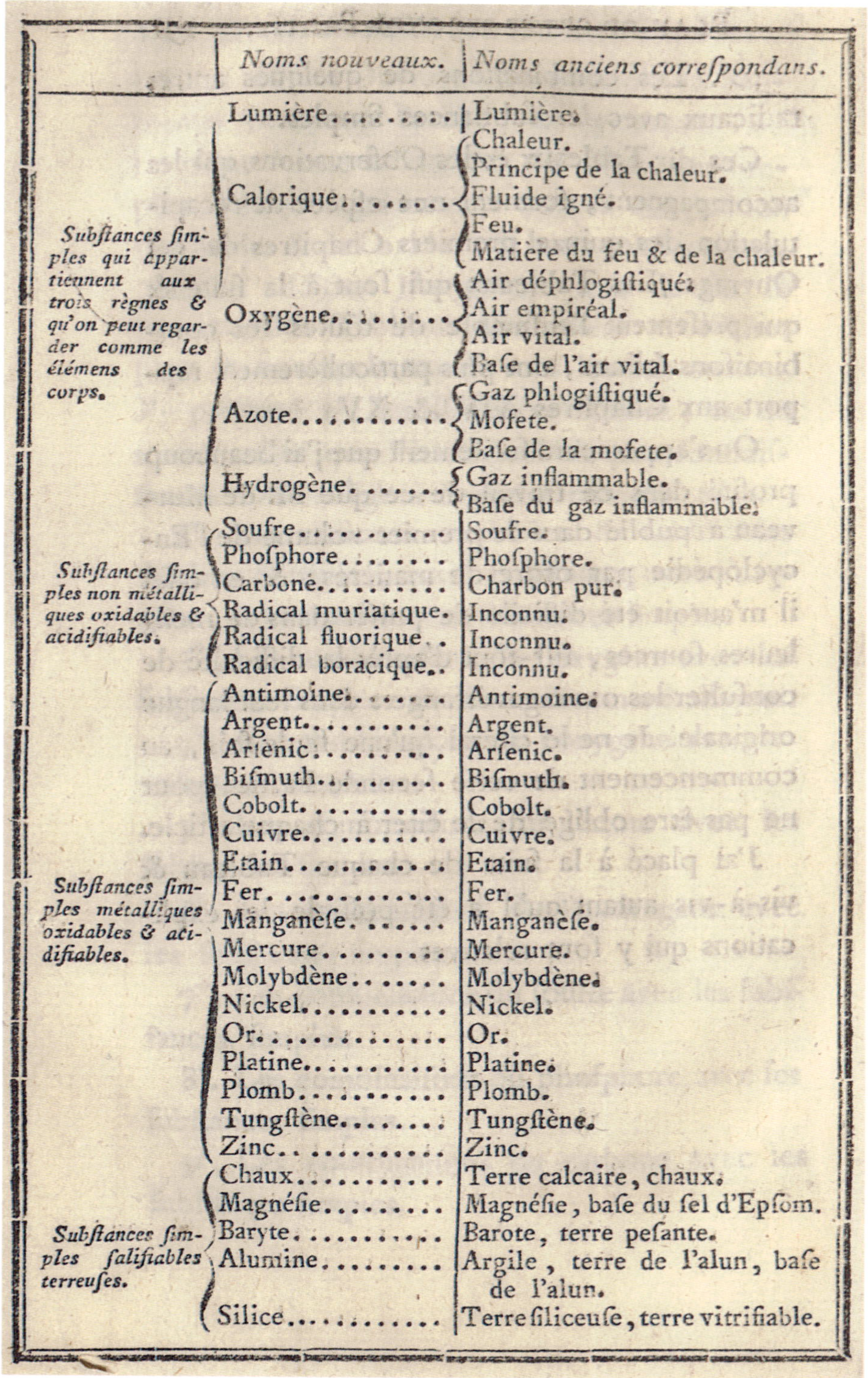

Noms nouveaux.	Noms anciens correspondans.
Lumière.................	Lumière.
Calorique............	Chaleur.
	Principe de la chaleur.
	Fluide igné.
	Feu.
	Matière du feu & de la chaleur.
Oxygène............	Air déphlogistiqué.
	Air empiréal.
	Air vital.
	Base de l'air vital.
Azote...............	Gaz phlogistiqué.
	Mofete.
	Base de la mofete.
Hydrogène........	Gaz inflammable.
	Base du gaz inflammable.
Soufre.............	Soufre.
Phosphore........	Phosphore.
Carbone............	Charbon pur.
Radical muriatique.	Inconnu.
Radical fluorique..	Inconnu.
Radical boracique..	Inconnu.
Antimoine........	Antimoine.
Argent............	Argent.
Arsenic...........	Arsenic.
Bismuth...........	Bismuth.
Cobolt............	Cobolt.
Cuivre............	Cuivre.
Etain.............	Etain.
Fer...............	Fer.
Manganèse........	Manganèse.
Mercure..........	Mercure.
Molybdène........	Molybdène.
Nickel............	Nickel.
Or................	Or.
Platine...........	Platine.
Plomb............	Plomb.
Tungstène........	Tungstène.
Zinc..............	Zinc.
Chaux............	Terre calcaire, chaux.
Magnésie..........	Magnésie, base du sel d'Epsom.
Baryte...........	Barote, terre pesante.
Alumine..........	Argile, terre de l'alun, base de l'alun.
Silice.............	Terre siliceuse, terre vitrifiable.

Row groupings (left-hand brace labels):

Substances simples qui appartiennent aux trois règnes & qu'on peut regarder comme les élémens des corps. — Lumière, Calorique, Oxygène, Azote, Hydrogène.

Substances simples non métalliques oxidables & acidifiables. — Soufre, Phosphore, Carbone, Radical muriatique, Radical fluorique, Radical boracique.

Substances simples métalliques oxidables & acidifiables. — Antimoine, Argent, Arsenic, Bismuth, Cobolt, Cuivre, Etain, Fer, Manganèse, Mercure, Molybdène, Nickel, Or, Platine, Plomb, Tungstène, Zinc.

Substances simples salifiables terreuses. — Chaux, Magnésie, Baryte, Alumine, Silice.

RIGHT: Antoine Lavoisier's chemical elements Table of Simple Substances. From his *Traité Élémentaire de Chimie*, Paris: Chez Cuchet, 1789, Library of Congress, Rare Book & Special Collections Division, Washington, DC.

prevailing theory at that time of how things burn, which ascribed this property to the putative element phlogiston. Chemists thought that phlogiston was released into the air during combustion; the more phlogiston a substance contained, the more combustible it was. Cavendish and some other chemists suspected that hydrogen might itself be pure phlogiston.

Cavendish was curious about what happened when inflammable air was burnt—a process thought to "phlogisticate" ordinary ("common") air. In 1781, he found that this process produced water, which would condense as droplets on the walls of a vessel in which the combustion took place. He wasn't the first to notice that, but he drew a striking conclusion: the water was being formed from common air and inflammable air. He was right, although his adherence to the phlogiston theory prevented him

BELOW: English scientist Henry Cavendish. Aquatint by C. Rosenberg after W. Alexander, nineteenth century, Wellcome Collection, London.

from expressing it in the way we would today: hydrogen combines with oxygen in the air to make water. Those latter two element names were coined by Antoine Lavoisier, who proposed this new view in the 1780s. Lavoisier's term for inflammable air, "hydrogen," means "generator of water." The implication was that water itself is not an element, but a compound.

Many chemists, especially in England, were outraged by Lavoisier's idea, which sought to replace the old phlogiston theory with these new elements hydrogen and oxygen. Lavoisier was right, but his victory didn't come without a battle.

Regardless of those controversies, hydrogen (or whatever you called it) was useful stuff. Because it is lighter than air, a balloon filled with it is buoyant and will rise up. In 1783 French chemist Jacques Charles filled a large balloon with hydrogen that bore him and his assistant Nicolas-Louis Robert aloft over Paris. It was not quite the first balloon flight, for Charles's compatriots, the brothers Joseph-Michel and Jacques-Étienne Montgolfier, had become airborne just a few days before in a hot-air balloon (where the buoyancy arises because warm air is less dense than cool air).

Ballooning became a sensation in the late eighteenth century, for the first time allowing people to see the land from the sky. From the mid-nineteenth century, hydrogen balloons were fitted with steam-powered and then electric engines that drove propellers, making them a steerable form of transport—for leisure, transport, and war. But hydrogen's inflammability was a constant danger, and the golden age of the airship came to an end, after regular transatlantic and even transglobal flights in the early twentieth century, with the Hindenburg disaster in New Jersey, USA, in 1937.

OPPOSITE: Filling a balloon with hot air. From Barthélemy Faujas-de-St.-Fond's *Description des Expériences de la Machine Aérostatique de MM. de Montgolfier*, Paris: Chez Cuchet, 1783, Science History Institute, Philadelphia.

OXYGEN

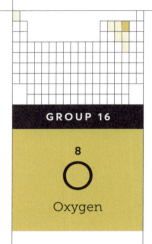

GROUP 16

8

O

Oxygen

Non-metal

ATOMIC NUMBER
8

ATOMIC WEIGHT
15.999

PHASE AT STP
Gas

OPPOSITE: Jacques-Louis David's oil on canvas *Antoine-Laurent Lavoisier and His Wife* (Marie-Anne-Pierrette Paulze), 1788. Purchase, Mr. and Mrs. Charles Wrightsman Gift, in honor of Everett Fahy, 1977, The Metropolitan Museum of Art, New York.

As Antoine Lavoisier's new chemical terminology spread, so did his ideas—because it was hard to use those terms without implicitly accepting the ideas they represented. Central to his revision of chemistry was the element that he named *oxygène*, which means "maker of acids" (Lavoisier thought, wrongly, that all acids contain oxygen). This substance is a gas at normal temperatures and pressures, and constitutes one-fifth of the air. Lavoisier was certainly not the first to make and identify it in a chemical process, but he was the first to recognize that it is indeed an element in its own right.

The second half of the eighteenth century is often regarded as the age of "pneumatic chemistry," meaning the study of gases—or what the chemists of that time called "airs." It had long been known that various chemical processes emitted "airs," which could often be seen bubbling from the reaction. But only from this time did chemists begin to make clear distinctions between them, thanks in part to the invention of apparatus for collecting these gases. Henry Cavendish's study of what became known as hydrogen was a part of this emerging tradition.

One of the preeminent pneumatic chemists was the Englishman Joseph Priestley, a religious nonconformist and political radical, who identified around twenty different types of "air"—among them, the compounds we now call ammonia, nitric oxide, and hydrogen chloride. Like many of his contemporaries he regarded these as varieties of common air in different degrees of purity or contamination. He was a firm believer in the phlogiston theory of combustion.

In 1774 Priestley collected an "air" produced by heating red mercury oxide—an experiment that had certainly been done previously by the French pharmacist Pierre Bayen and probably by earlier chymists and alchemists. Priestley found that when a burning candle was placed in a vessel filled with this gas, it burned even more brightly, while a lump of burning charcoal glowed with incandescence. This propensity to make things burn better, he decided, must be because the "air" had less phlogiston and so was more able to soak it up: he called it "dephlogisticated air." He found too that mice placed in a glass vessel filled with the new air could keep breathing for longer than in a vessel of common air. This emboldened Priestley to try breathing it himself: "my breath," he wrote, "felt peculiarly light and easy for some time afterward."

Priestley was not the only one studying this marvelous air. In Sweden, the apothecary Carl Wilhelm Scheele had found around 1771–1772 that heating up the compound known as niter or saltpeter (potassium nitrate) released an "air" too. That experiment had been tried a hundred years earlier by Robert Boyle's assistant John Mayow, who reported that blood exposed to this air became a brighter red. Scheele found that this air also enhanced burning,

LEFT: Pneumatic trough and other equipment used in Joseph Priestley's various experiments on oxygen and other gases. From Priestley's *Experiments and Observations on Different Kinds of Air*, 1774–1786, London: Printed for J. Johnson, Wellcome Collection, London.

and he called it "fire air." Priestley had no knowledge of Scheele's work, however—the Swede did not publish it until 1777—and so he didn't make any connection between his dephlogisticated air and fire air.

These airs were, of course, both in fact oxygen—and it was Lavoisier who said so. Lavoisier already knew of Bayen's studies when Priestley came to dine with him in Paris in October 1774 and discussed his own findings. Later Priestley sent the Frenchman a sample of his gas, which Lavoisier decided was an especially "pure" or "true" air. Scheele also sent Lavoisier a letter in late 1774 describing his fire air.

Finally, Lavoisier put it all together. Common air, he decided, contains only around a quarter (actually closer to a fifth) of the more fundamental substance he called "true air"—and in 1777 he pronounced it a new element: oxygen. It is this, Lavoisier said, and not phlogiston that is the real begetter of combustion. When substances burn, they do not release phlogiston; rather, they combine with oxygen from the air. That is why metals get heavier when heated in air to form what chemists call a

calx: it is the metal oxide. When Cavendish's inflammable air burns to form water, this is in fact the reaction of hydrogen with oxygen.

Whether, then, it is Priestley, Scheele or Lavoisier who merits being called the discoverer of oxygen has been much disputed. In their own time that argument was conducted with some rancor—partly because Lavoisier was ungenerous about acknowledging the prior work of the other two, and partly because of nationalistic rivalry between England and France. Today most historians would see the argument as moot; like many scientific discoveries, this one did not happen all at once. Lavoisier certainly deserves to be recognized, however, as the man who made sense of a mass of confusing and sometimes conflicting experimental results about combustion, calx formation, and respiration with the single unifying idea that there is an element in the air called oxygen.

OPPOSITE: "Experiments on the Air Pump." Engraving from William Henry Hall's *The New Royal Encyclopaedia*, London: Charles Cooke, 1795, Second edition, Vol. 1, Wellcome Collection, London.

NITROGEN

GROUP 15

7
N
Nitrogen

Non-metal

ATOMIC NUMBER
7

ATOMIC WEIGHT
14.007

PHASE AT STP
Gas

Air was terribly confusing. We need it to survive—and yet there seemed, to the young pneumatic chemist Daniel Rutherford, working at the University of Edinburgh in the 1770s, to be something noxious in it too, capable of extinguishing life.

Rutherford's mentor Joseph Black showed in the 1750s that when carbonate salts such as lime (calcium carbonate) are heated or treated with acid, they emit a gas that puts out candle flames and kills animals that breathe it in. Black called this "fixed air," because it could seemingly be "fixed" into such carbonates—for example, by reacting the gas with quicklime (calcium oxide). Black showed that fixed air was also present in our exhalation, and so was apparently a product of respiration.

In his thesis for a medical doctorate in 1772, Rutherford called this gas "mephitic air," from the Greek name given to a legendary noxious emission. If a small animal were kept in a closed vessel full of air, it would eventually die of asphyxiation, even though some "air" still remained in the vessel. And chemical analysis showed that some of this "air" was indeed mephitic air, so that all seemed to make sense. Except that even if Rutherford removed the

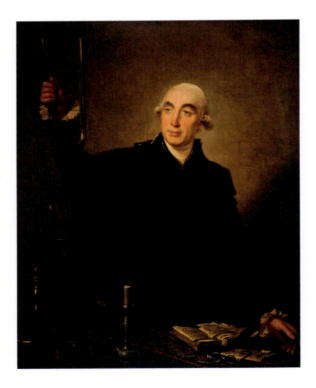

LEFT: David Martin's portrait *Professor Joseph Black, 1728–1799, Chemist*, 1787, Scottish National Portrait Gallery, Glasgow, Scotland (Private Collection on long-term loan to the National Galleries of Scotland).

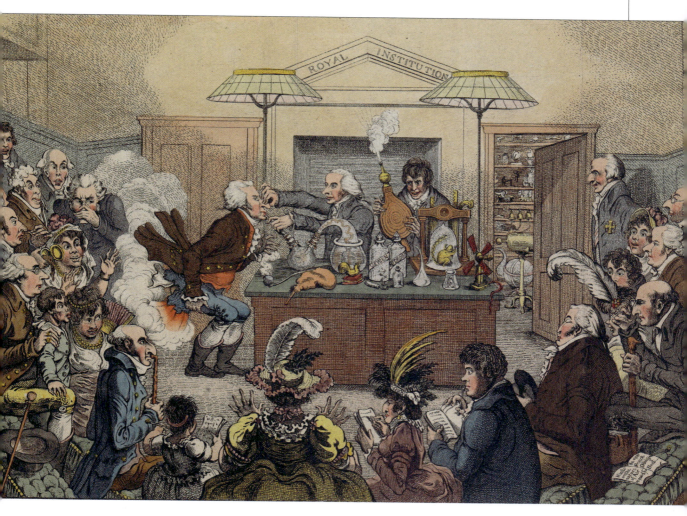

ABOVE: A lecture on pneumatics at London's Royal Institution in which Thomas Young demonstrates the effects of nitrous oxides on Sir J. C. Hippisley, while Sir Humphrey Davy holds bellows. Among the audience are Count Rumford and Lord Stanhope. James Gillray's colored etching *Scientific Researches!— New Discoveries in Pneumaticks! —or—an Experimental Lecture on the Powers of Air*, Wellcome Collection, London.

mephitic gas with quicklime, animals would still die if they were left to breathe the "air" that remained. In other words, there was another "noxious air" in there, too. Both Joseph Priestley and Henry Cavendish had also shown previously that when a candle was burned in air, it decreased in volume by about a fifth: some was consumed by the flame, but some remained.

If we were to persist with Rutherford and Black's way of thinking about these observations, couched in the language of phlogiston, it would get even more confusing. So let's instead jump ahead to the chemical language of a few decades later to make sense of it. Fixed air is carbon dioxide, which is indeed produced in our lungs when we exhale. But Rutherford's second "noxious air" is nitrogen gas:

the other component of air, of which it makes up about four-fifths. Nitrogen is not really poisonous at all—we would clearly be in trouble if it was—but neither will it support life. What we need from air is oxygen; nitrogen is just a kind of inert background gas. Exposed to an atmosphere of pure nitrogen, like Rutherford's mice, we will die simply from a lack of oxygen.

By removing oxygen and carbon dioxide from air, Rutherford was left with almost pure nitrogen (there are tiny amounts of other gaseous elements there too, as others were later to discover). And so he is credited with the discovery of nitrogen, even though he never called it that, nor really showed any awareness that it was a true element.

The name means "maker of niter"—for that substance (potassium nitrate), long known as a component of gunpowder, contains it. Antoine Lavoisier (who knew of Rutherford's work) discerned in the 1780s that common air is, in fact, a mixture of two gases—the "highly respirable air" that he came to call oxygen, and a second, more abundant gas that—apologies for the confusion—he called by the same name that Rutherford had used for Black's "fixed air": mephitic air (because it was unbreathable). What Lavoisier meant by that was nitrogen.

BELOW: A study of nitrogen in plants. From Théodore de Saussure's *Recherches Chimiques Sur la Végétation*, Paris: Nyon, 1804, Science History Institute, Philadelphia.

When Lavoisier began to abandon this terminology of "airs" in favor of new element names, he chose to denote his mephitic air as *azot*, from the Greek word indicating that it was incompatible with life. He showed with further chemical experiments that it is a part of nitric acid, from which saltpeter can be made. He admitted that for this reason *nitrogen* would be a good name too: but he stuck with *azot*, and to this day the element nitrogen is still called *azote* in French. In England, the alternative caught on—well, in those days the English were never keen to do what the French did.

Far-reaching uses

Nitrogen gas is highly inert because it contains molecules in which two nitrogen atoms are tightly bound together—by what chemists call a triple bond, which takes a lot of energy to break. All the same, nitrogen is one of the most abundant elements in living matter, where among other things it is present in the amino acids of which proteins are

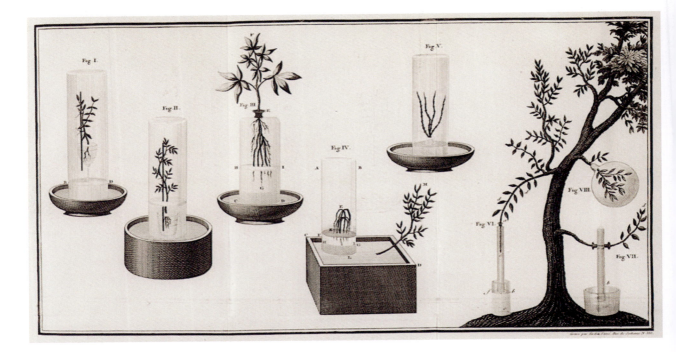

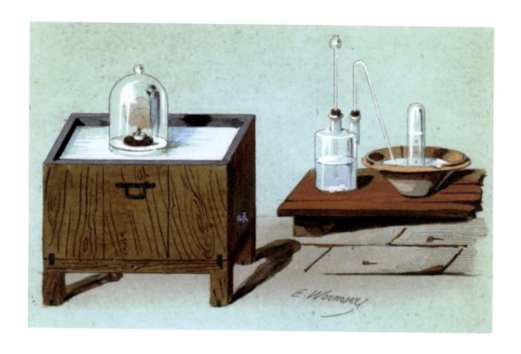

ABOVE: Preparation of nitrogen (left) and hydrogen (right). From J. Pelouze et E. Fremy's atlas *Notions Générales de Chimie*, Paris: Victor Masson, 1853, Plate II, National Central Library of Florence.

composed, and in DNA too. Getting unreactive nitrogen out of the air and into living organisms is no mean feat: it is largely done by specialized soil microorganisms called diazotrophs—many of them bacteria—that exist in a symbiotic relationship with plants. They contain enzymes called nitrogenases that can split apart the doughty triple bond using ingenious chemistry.

As it is an essential nutrient for plant growth, nitrogen is a key component of fertilizers—this is one of the uses of saltpeter. But making chemical fertilizers from atmospheric nitrogen means replicating the ability of diazotrophs to "fix" it from the gaseous form into more reactive compounds, usually starting with ammonia (a compound of nitrogen and hydrogen). That has been done since the early twentieth century using an industrial process called the Haber-Bosch process, involving a metal (iron) catalyst. By supplying

the fertilizer essential for boosting crop yields, the Haber-Bosch fixation of nitrogen is perhaps the most important enabler—for better or worse—of the massive population growth in the twentieth century.

Many explosives, from dynamite and TNT to Semtex, are also compounds of nitrogen. This is because reactions that recombine nitrogen atoms into two-atom molecules, with their strong and stable chemical bond, release a large amount of energy. The flipside of nitrogen gas's inertness, you might say, is the extreme and even hazardous reactivity of some of nitrogen's compounds.

CARBON

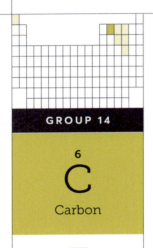

GROUP 14

6

C

Carbon

Non-metal

ATOMIC NUMBER
6

ATOMIC WEIGHT
12.011

PHASE AT STP
Solid

The "discovery" of carbon seems an almost absurd concept. It is the basis of all life on Earth. It is, as soot and charcoal, the product of the fire that warmed and nurtured *Homo sapiens* in the dimmest reaches of prehistory. In the form of the minerals graphite and diamond, it has always been around us in pure form—in the case of the latter, as a form of alluring and dazzling brilliance. What could the discovery of carbon even mean?

People have long produced and dealt in charcoal, which was used as a fuel for fires (it burns better than wood) and as a black pigment for cave painting. Later it was found that charcoal will "reduce" some metal ores to the pure metal: the carbon strips away the oxygen from oxide minerals of copper, tin, and iron. It was a component of gunpowder, and was sometimes used in medicines or as a preservative: water might be stored in charred barrels to keep it tasting good, as charcoal is a good absorber of impurities and microorganisms (that's why it is still used in filtration today).

Diamonds, formed from carbon-rich fluids circulating deep in the Earth and sometimes brought to the surface by volcanic activity, were known since ancient times—the Greeks called them *adamas*, meaning "invincible," because of their hardness and durability. Diamonds were mined and traded from around the first millennium BC in India, which was essentially the world's only source of diamonds until the eighteenth century. The famous 105.6-carat Koh-i-Noor diamond in the crown of the British monarch came from there, part of the plunder of empire during the Victorian age.

Carbon's other mineral form, graphite, has also been mined for many centuries: it was used in ancient times as a pigment, but demand surged from around the sixteenth century when its softness made it handy as a lubricant for molds used to cast cannonballs, and later to ensure the smooth running of machinery, as well as for use in pencils.

Same element, different substance

The contrast between graphite and diamond supplies the most dramatic indication of how the properties of an element are defined not so much by the nature of its atoms as by how they are joined together. The pattern of chemical bonds between carbon atoms is very different in the two substances: in diamond, it joins them into a three-dimensional crystalline framework of immense strength and perfect transparency, while in graphite the atoms are linked into sheets of hexagonal rings that can slide over one another and which make the material absorb virtually all visible light that falls on it. Coal is similar to graphite, being formed in the Earth from organic material (largely the abundant vegetation of the Carboniferous period) that has been squeezed and heated until it becomes mostly graphite-like carbon.

ABOVE: French chemist Henri Moissan trying to make diamonds, ca. 1890s, Bain News Service, Library of Congress Prints & Photographs Division, Washington, DC.

Because of these marked differences between graphite and diamond, it's not surprising that chemists took a long time to appreciate that diamond and graphite/coal/charcoal were both composed of the same single element. Diamond was particularly difficult to study, not just because it was so costly but because its very robustness made it hard to analyze—literally, to "break down" so that its ingredients could be examined. In 1694 two experimenters in Florence named Giuseppe Averani and Cipriano Targioni used lenses to focus sunlight onto diamonds and showed that they could be vaporized by the heat—an expensive and rather alarming experiment sponsored by the Grand Duke of Tuscany. Almost a hundred years later the French chemist Pierre Macquer and some of his collaborators repeated the experiment, showing not only that diamond could be burned to nothing but

that in some circumstances it was transformed into a charcoal-like material, for which the French word was *charbone*.

Hearing of those studies, Antoine Lavoisier took up the challenge in the early 1770s using a gigantic lens about a meter in diameter. It was Lavoisier's great insight that he thought about the process not merely as a kind of evaporation of the gem, but as a chemical reaction between the diamond and oxygen in the air. He showed that the gem was burned into a gas that he identified as Joseph Black's "fixed air" (see page 114)—that is, carbon dioxide, which is also produced by burning *charbone*. What else could this mean, but that diamond and charcoal are one and the same: an element that came to bear the name of the latter?

Still, that was such a strange and surprising conclusion that it was not clearly proved until the end of the century, after Lavoisier had gone to the guillotine. In December 1796, the English chemist Smithson Tennant read to the Royal Society a paper called "On the Nature of the Diamond." Although Lavoisier had noted the resemblance of charcoal to

ABOVE: A graphite mine in Batougal, Saiansk, eastern Siberia. From Louis Simonin's *La Vie Souterraine ou, Les Mines and Les Mineurs*, 1868, Science Photo Library, London.

OPPOSITE: Diamonds and corundum. From Max Hermann Bauer's book on geology, *Edelsteinkunde*, Leipzig: C. H. Tauchnitz, 1909, Plate 1, University of Chicago.

diamond, he said, the Frenchman had concluded no more than that "each of those substances belonged to the class of inflammable bodies." Tennant reported his careful experiments in which he measured that the amount of fixed air produced from heating a diamond until it disappeared was the same as would be produced from burning an equal mass of charcoal. They are the same.

If this was so, might one be able to convert lowly charcoal (or graphite) into valuable diamond? That prospect enticed chemists throughout the following

century, who supposed that it might be done by heating charcoal or graphite while subjecting it to great pressure. In 1893 the French chemist Henri Moissan claimed to have succeeded, but that now seems deeply improbable. The first convincing artificial synthesis of diamond from graphite-like carbon was only reported in 1955, when researchers at the General Electric Company in Schenectady, New York, achieved it using a 400-ton hydraulic press to produce pressures a hundred thousand times greater than normal atmospheric pressure.

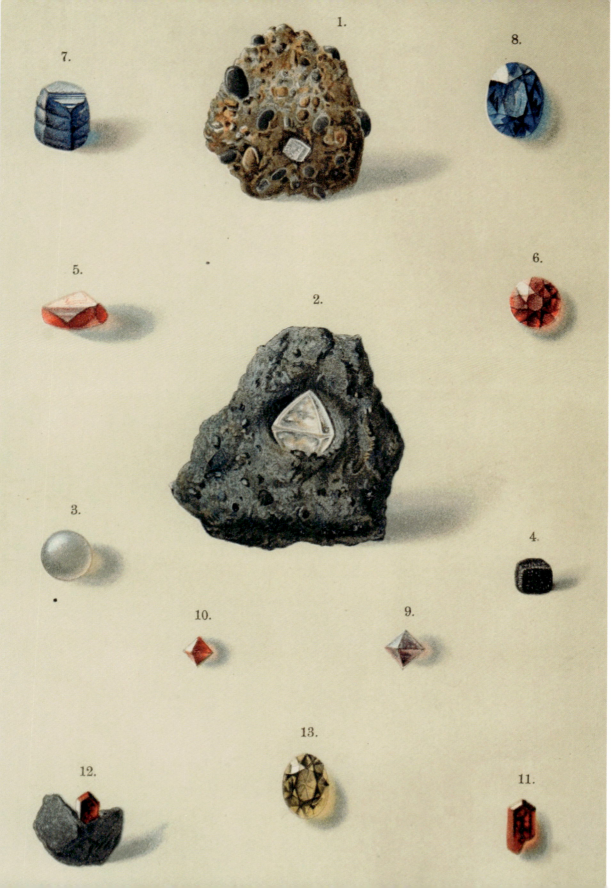

CALORIC

By the end of the eighteenth century, scientists had figured out the true chemical nature of three of the classical elements: earth, air, and water. Air was, in Antoine Lavoisier's scheme, a mixture of the elemental gases oxygen and nitrogen, while water was a compound formed by the reaction of hydrogen and oxygen. There were many "earths"—rocks and minerals of all descriptions, within which familiar and new elements were steadily being identified.

But what of the fourth classical element, fire? This is not so much a substance as a process, and it seemed to be one of great complexity. There is substance in the flame, for out of it (if we are speaking of a candle or a log fire) comes soot and carbon dioxide. But it produces light too. And perhaps most importantly of all for most purposes, a fire produces heat. It seemed reasonable to suppose that much of the answer to the question "What is fire?" might be found in the answer to another: "What is heat?"

Heat is not uniquely produced by fire, of course. You can create it by rubbing your hands together. Our bodies seem to generate it—they are typically warmer than their surroundings. Many chemical reactions will generate heat, without any flame involved. It will come too from the flow of electricity: the snap of a spark can burn you, while a lightning bolt may do far worse.

There was a clear sense in which heat seems to *flow*. Place one end of an iron rod into a fire, and before long the other end becomes too hot to hold, as heat passes along the rod. Heat streams out from a flame. And we are familiar with substances that flow—with *fluids*. Water flows in streams and rivers, while gases too were recognized as fluids: carbon dioxide from a candle flame may flow down a tube and be collected in a vessel. It seemed very reasonable to

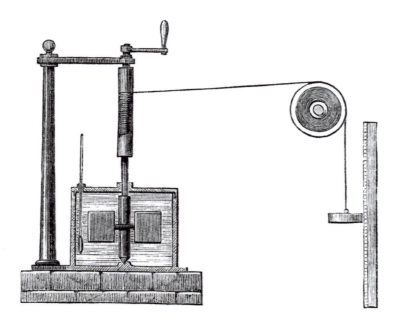

LEFT: James Prescott Joule's apparatus for measuring the mechanical equivalent of heat. Engraving from *Harper's New Monthly Magazine*, No. 231, August 1869.

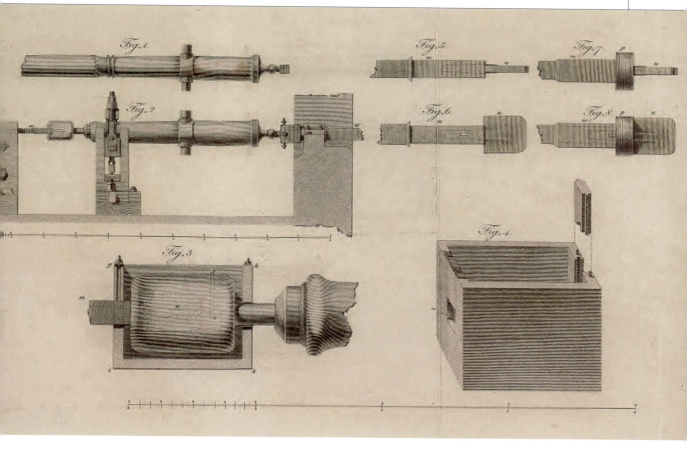

suppose, then, that heat was a sort of fluid too, albeit one so tenuous that it could pass through solid materials.

The same seemed true of coldness: put the iron bar into an ice bucket instead, and the coldness spreads along it. In ancient times, some philosophers imagined that heat and cold were opposed substances or tendencies, perhaps kinds of particles that were emitted from bodies.

The phlogiston theory, which dominated ideas about combustion throughout most of the eighteenth century, seemed to some to answer this puzzle: phlogiston was itself the *substance of heat*. When Antoine Lavoisier replaced that idea with his oxygen theory of combustion, he had to find some new way of accounting for heat. He didn't abandon the notion that it was a substance, but merely renamed it: the substance of heat, he said in 1783, was a "subtle fluid" that he called *calorique* (caloric). (Might there then also be a "substance of cold"—a

ABOVE: Benjamin Thompson, Count of Rumford's "An Inquiry concerning the Source of Heat which is excited by Friction." From *Philosophical Transactions of the Royal Society of London for the Year MDCCXCVIII* (1798), Part I, Vol. 88, Natural History Museum Library, London.

frigoric? Some thought so, but others said cold was merely the absence of caloric.) In his great work of 1789, *Traité Élémentaire de Chimie*, Lavoisier included caloric in his list of thirty-three known elements.

As with phlogiston and aether, including caloric in a book on the discovery of the elements might cause some raised eyebrows among chemists, who know that this too is another "element that never was." But we don't get a proper understanding of history if we weed from it all the ideas that we know now to be wrong. For one thing, we lose sight of the fact that scientists often only come to the right conclusions by leaning on notions that are later

ABOVE: Ice-calorimeter for measuring heat. From Antoine Lavoisier's *Traité Élémentaire de Chimie*, Paris: Chez Cuchet, 1789, Plate VI, Science History Institute, Philadelphia.

disproven: they are not errors or mistakes, but rather signposts along the road to a better understanding of the world. After all, caloric seemed to make sense of several observations during Lavoisier's time. When a gas gets warmer, for example, it expands—and what better explanation than that it absorbs some extra fluid like caloric? Lavoisier devised apparatus for measuring the "flow of caloric" between substances, a technique called calorimetry, which is still the term used today for measuring changes in heat.

The birth of thermodynamics

When the French military engineer Sadi Carnot developed his theory of how engines driven by heat (such as the steam engine) work in the 1820s, he drew on the caloric theory of his compatriot Lavoisier, according to which this substance was transferred from hot bodies to cold ones. Carnot's work supplied the basis for the science of thermodynamics—literally, the motion of heat—which remains one of the central pillars of physical theory today. Not a bad legacy for a fictitious element.

Still, fictitious it is. In 1798 the British (American-born) scientist Benjamin Thompson (who had the title Count Rumford) published a very different account of what heat truly was. Lavoisier had supposed that caloric was a conserved substance: it was never created or destroyed, but merely flowed from one place to another. But Rumford described experiments he had conducted while superintending the boring of cannons in Germany. This process generated a lot of frictional

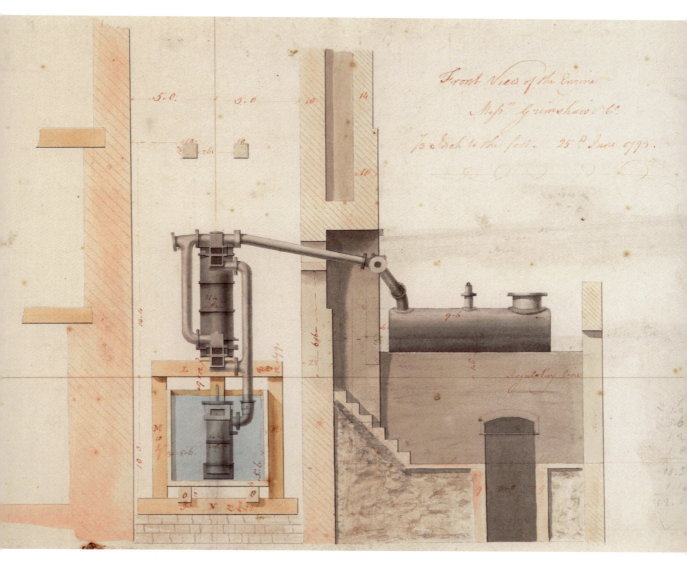

heat, and the hot brass needed to be cooled with water. Rumford showed that repeated boring operations could heat the water again and again, as if a supply of supposed caloric were inexhaustible.

Heat, he concluded, was not produced by a substance but by a process, namely *motion*. As to motion of what, and through what cause, he couldn't say. But researchers later in the nineteenth century, such as James Joule and James Clerk Maxwell, developed the idea that heat was a property arising

ABOVE: Steam engine designed by Matthew Boulton and James Watt for Messrs Grimshaw & Co., Sunderland, 1795, Institution of Mechanical Engineers, London.

from the movements of the invisibly small particles—the atoms and molecules—of which substances are composed. This so-called "kinetic theory" of heat (meaning that it relates to motion) became the foundation of the modern theory of thermodynamics.

CHLORINE

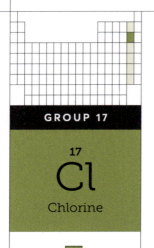

GROUP 17

17
Cl
Chlorine

Halogen

ATOMIC NUMBER
17

ATOMIC WEIGHT
35.45

PHASE AT STP
Gas

Of all the natural compounds containing chlorine, probably the most important for human use is sodium chloride: common salt. It can be extracted from seawater by letting the water evaporate, an ancient means of salt manufacture that is still practiced today. There are also large mineral deposits of common salt, created by the evaporation of briny seas in the geological past.

In salt, the chlorine and sodium atoms are tightly bound up with one another, and this substance, vital for human health, gives no hint of the toxic, noxious nature of the element hidden within it. The recognition of chlorine as a distinct element owes much more to another kind of salt, known in ancient times as sal ammoniac, which is ammonium chloride. This can occur naturally as a white mineral in volcanic regions, but it's rare. Sal ammoniac only truly entered the chemist's repertoire through the investigations of the Arabic alchemists around the ninth century, who reported that it could be obtained by burning camel dung. The Persian alchemist and physician Muḥammad ibn Zakariyyā al-Rāzī described it in the tenth century, and explained how it can be distilled. When heated, it decomposes into the pungent gases ammonia and hydrogen chloride.

That was an important discovery, because hydrogen chloride dissolves readily in water to form hydrochloric acid (sometimes still known today by the alchemical name spirit of salts), which along with sulfuric and nitric acids is one of the so-called mineral acids that were among the chemist's most potent reagents. We saw earlier how a mixture of hydrochloric and nitric acids could

BELOW: Western depiction of Muḥammad ibn Zakariyyā al-Rāzī. From a translation of his *Kitab al-Hawi fi al-tibb* (The Comprehensive Book on Medicine), 1529, Qatar National Library.

even dissolve that most unreactive and "regal" of metals, gold—for which reason this mixture was awarded the name *aqua regia* (water of kings). This marvelous solvent was not, however, made by mixing the two pure acids together, but instead by dissolving sal ammoniac in nitric acid.

It's not known exactly when pure hydrochloric acid was first made. The Arabic scholars clearly knew enough to make it, but whether they did so is another matter. Probably it was made by accident several times before it first starts to be clearly described around the sixteenth and seventeenth centuries, when chemists such as the Germans Andreas Libavius and Johann Rudolph Glauber give recipes for it. Libavius made it by distilling common salt with pieces of clay. It was often known to early chemists as muriatic acid, meaning "marine acid."

Isolating chlorine

It wasn't until the late eighteenth century that chemists figured out how to get elemental chlorine out of this acid. Carl Wilhelm Scheele tried heating muriatic acid with the mineral form of manganese oxide, pyrolusite, and found that it gave off a dense greenish gas that must have alarmed him: he says it has a choking smell that is "highly oppressive to the lungs." Scheele found that the gas would react with most metals, coating them with a colored patina (the metal chloride), would dissolve in water to make an acid, and bleach the color from flowers and leaves.

This stuff soon became used, dissolved in water, as a bleach in the textiles industry, making it a far faster process than the traditional method of sun-bleaching. In 1785 the French chemist Claude Berthollet showed that a better bleach could be made by dissolving the gas in sodium hydroxide solution, which forms sodium hypochlorite—the household bleach still in use today (and which has a strong whiff of chlorine itself).

As far as Scheele was concerned, this gas was some kind of compound; cleaving to the phlogiston theory, he called it dephlogisticated muriatic acid. (This makes sense when decoded: phlogiston could be confused with hydrogen, and so if you "dephlogisticate" hydrogen chloride, you're left

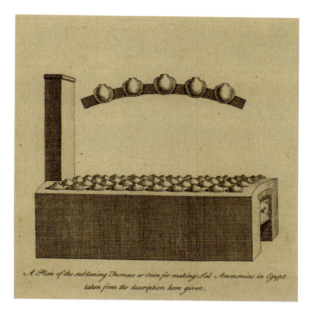

A Plan of the subliming Furnace or oven for making Sal Ammoniac in Egypt. taken from the description here given.

ABOVE: C. Linnaeus and John Ellis's "A Plan…for making Sal Ammoniac in Egypt." From *Philosophical Transactions of the Royal Society* (1683–1775), 1759, Vol. 51, Table XI, Royal Society of London.

with chlorine.) Some contemporaries of Scheele, including Berthollet, believed that his acrid gas contained a hitherto unknown element, muriaticum, combined with oxygen. But when in 1809 the Frenchmen Joseph Louis Gay-Lussac and Louis-Jacques Thénard tried to react it with charcoal to remove the oxygen from what, in Lavoisier's scheme, was now called "oxymuriatic acid," they saw no change. Could it perhaps be an element in itself?

The British chemist Humphry Davy took that idea seriously. When in 1810 he tried the same thing and found the same result, he declared that this gas was indeed an element, and proposed a name: chlorine, from the Greek word *chloros* meaning "pale yellow-green." (The green plant pigment chlorophyll has the same etymological root, though it contains no chlorine.) Chlorine gas will liquefy at -29°F, but in 1823 Davy's protégé Michael Faraday "took advantage of the late cold weather" to produce a sample of liquid chlorine—a yellow liquid—without having to resort quite to those frigid extremes.

FLUORINE, IODINE, AND BROMINE

GROUP 17	
9	F
Fluorine	Gas

HALOGEN

ATOMIC WEIGHT: 18.998

GROUP 17	
53	I
Iodine	Solid

HALOGEN

ATOMIC WEIGHT: 126.90

GROUP 17	
35	Br
Bromine	Liquid

HALOGEN

ATOMIC WEIGHT: 79.904

Another mineral that drew the voracious attention of Carl Wilhelm Scheele was called fluorite, its name derived from the Latin *fluoere*: "to flow," because it would melt at lower temperatures than many other minerals. "The fire melts [*fluores*]," says Agricola's narrator Bermannus in his sixteenth-century book of mineralogy, "and makes them as fluid as ice in the sun."

Fluorite has the unusual property of glowing when heated: the property that gives rise to the word fluorescence. It does this because the mineral stores energy absorbed from high-energy radiation, such as natural radioactivity in the Earth or cosmic rays from space, locking it up in "defects" where the perfect atomic ordering of the crystal is disrupted. The energy can only be released (as light) when the crystal is loosened up by heat. Scheele did not and could not understand any of that, but the fluorescence of fluorite fascinated him.

Scheele found that the mineral would release an acidic gas when heated strongly, which he called fluor spar acid. This was also known as Swedish acid or Sparry acid—but in the 1780s the French chemists in Antoine Lavoisier's circle sought to reform and rationalize chemical nomenclature, and they proposed calling it fluoric acid.

LEFT: Swedish chemist Jöns Jacob Berzelius. Lithograph by J. C. Formentin after J. V. C. Way, 1826, Wellcome Collection, London.

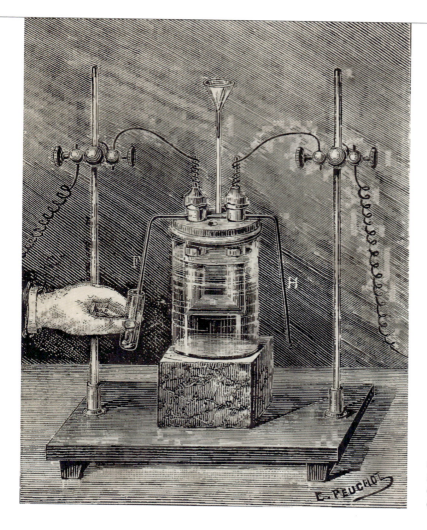

E. PEUGHOT

LEFT: French chemist Henri Moissan's *Recherches sur l'Isolement du Fluor*, Paris: Gauthier-Villars, 1887, Francis A. Countway Library of Medicine, Harvard.

What was it? Lavoisier suspected that, like muriatic acid, it was a combination of some element with oxygen. But when Humphry Davy showed that there was no oxygen in the former—that it contained the element chlorine instead—the French scientist André-Marie Ampère wrote to him suggesting that perhaps fluoric acid was similar. How about, Ampère said, giving the hypothetical new element in this acid the name "fluorine"? Davy liked that idea, and so it was.

The seaweed elements

A year later, in 1813, Humphry Davy went to visit Ampère in Paris—the Napoleonic Wars were raging, but eminent chemists such as Davy were considered by Napoleon to be exempt from hostilities. There, Ampère gave Davy a sample of a substance that had been isolated from seaweed by the French chemist Bernard Courtois in 1811. Courtois had been making alkali from the ashes of the seaweed, and had found that by treating it with sulfuric acid he could obtain a strange, smelly violet vapor that might be condensed into crystals with a dark, metallic sheen like graphite. French chemists found this stuff would combine with hydrogen to make an acid similar to muriatic acid, suggesting that it was also a new element similar to chlorine. The French researchers called this *ione*, after the Greek for "violet"; but Davy preferred to modify the name—making it closer to chlorine and fluorine—as *iodine*.

Fig. 1
Experiment I.

Chromatic Equivalents.
Fig. 2. Exp. XXVIII.

Definitive or Fundamental Scale of Colours
Fig. 3

A convenient test was soon found for the presence of iodine: it turns a solution of starch dark blue. While studying a sample of it obtained from a Mediterranean seaweed in 1825, French pharmacist Antoine-Jerôme Balard found that underneath the blue layer of starch-iodine in his flask there was a layer of another fluid, deep yellow-orange in color. At first he called this stuff muride (from the same briny root as muriatic acid); but then, noting it had a head-turning pungency, he proposed the name *brome*, from the Greek for "stench." An English textbook of 1827 suggested the name should, in that language, become *bromine*.

Partly because "fluoric acid"—what we now call hydrofluoric acid—was so corrosive and difficult to work with, pure fluorine wasn't isolated until 1886,

by French chemist Henri Moissan: an achievement that won him the Nobel Prize in Chemistry in 1906. Fluorine is, if anything, even worse than its acid, being highly toxic and one of the most reactive substances known to chemistry. The four elements fluorine, chlorine, bromine, and iodine all sit within the same column of the periodic table, and are known collectively as the halogens, meaning "salt-makers"—for all of them react with metals to make salts, of which that composed of sodium and chlorine is common table salt, the predominant source of the oceans' briny tang. The collective name was first proposed in 1811 for chlorine alone, but was rejected. In 1826 the Swede Jöns Jakob Berzelius, a vigorous reformer of chemical nomenclature, recycled it once it became clear that chlorine, iodine, and fluorine were three of a kind.

OPPOSITE: Frontispiece to George Field's *Chromatography*, London: Charles Tilt, 1835, Linda Hall Library of Science, Engineering, and Technology, Missouri. The book included details of all the new pigments, such as iodine scarlet, a new pigment of a most vivid and beautiful scarlet color.

CHROMIUM AND CADMIUM

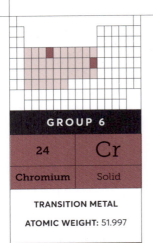

GROUP 6	
24	Cr
Chromium	Solid

TRANSITION METAL
ATOMIC WEIGHT: 51.997

GROUP 12	
48	Cd
Cadmium	Solid

TRANSITION METAL
ATOMIC WEIGHT: 112.41

In 1761 the German mineralogist Johann Gottlob Lehmann became a professor of chemistry at the Imperial Museum in St. Petersburg, and he began to examine the minerals of Russia. At a mine in the Ural mountains he came across a bright red mineral that he called *Rotbelierz*, meaning "red lead ore," an allusion to the ancient pigment red lead. It soon became used as a red pigment in paints, and gained the nickname Siberian red lead; later it was given the proper mineral name crocoite.

Crocoite was indeed an ore of lead, but what else did it contain? The French chemist Nicolas Louis Vauquelin (who had recently discovered the element beryllium in the mineral beryl) set out to answer the question after receiving a sample of the mineral in 1794. He found that reacting it with hydrochloric acid (as he'd have said, muriatic acid) produced a green substance, and Vauquelin used the standard procedure for extracting a metal from such a compound: heating it with charcoal. Sure enough, it produced

LEFT: French chemist Nicolas Louis Vauquelin. Engraving by F. J. Dequevauviller after C. J. Besselièvre, 1824, Wellcome Library, London.

ABOVE: The green pigment viridian used in the foreground of Jean-Baptiste-Camille Corot's oil on paper (laid on canvas) *The Roman Campagna, with the Claudian Aqueduct,* probably 1826, National Gallery, London.

a metal that was reported to be "grey, very hard, brittle, and easily crystallizes in small needles."

Vauquelin found that several of the compounds containing this metal, made from chemical treatments of crocoite, were strongly colored. He dissolved the powdered mineral in an alkali (potassium carbonate, as we'd now call it), and then neutralized it with nitric acid to produce a solution colored bright orange. When Vauquelin crystallized the dissolved salt, it was a deep, rich yellow. By varying the reaction conditions—for example, by adding lead sulfate—he could adjust the color of the product from a primrose yellow to an orange. The latter was the first pure orange pigment that painters had at their disposal aside from the expensive and poisonous realgar (see page 88), which is an arsenic compound.

In the light of these chromatic riches, the French chemist proposed that the metal be named after the Greek word for color itself: in French, *chrome,* which soon became standardized to chromium. Siberian red lead is a mineral form of lead chromate; the pigment that became known as chrome yellow is

a synthetic version. The German chemist Martin Klaproth independently discovered chromium in crocoite a year after Vauquelin.

This new metal was a boon to the paint industry, especially after a different chromium-containing ore (iron chromate) was discovered in the USA in 1808, in France in 1818, and in the British Shetland Islands two years after that. Even though pure chrome yellow remained quite expensive in the early nineteenth century, it was such a strong color that it could be diluted with cheap white "extenders" such as barium sulfate to make an inexpensive canary-yellow paint, which became popular for the coaches used for transport throughout Europe.

The green compound that Vauquelin made simply by roasting crocoite in air was also the source of a pigment. "On account of the beautiful emerald colour which it communicates," Vauquelin

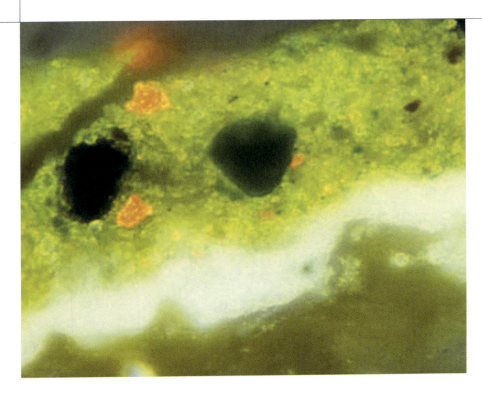

LEFT: The green pigment viridian shown in a cross section of Jean-Baptiste-Camille Corot's *The Roman Campagna, with the Claudian Aqueduct*, probably 1826, National Gallery, London.

wrote, it "will furnish painters in enamel with the means of enriching their pictures." In 1838 the Parisian color-maker Antoine-Claude Pannetier found out how to make this green purer and more bluish, creating the popular pigment that became known as viridian.

Out of a yellow chimney

As pigment manufacture became big business in the nineteenth century, chemists were constantly on the lookout for new materials that might be used in this way. So, when the German chemist Friedrich Stromeyer noticed in 1817 that a yellow-colored substance was deposited in the chimneys of a zinc smelting factory at Salzgitter, in Saxony, he decided to investigate further. By "reducing" this material using charcoal in the usual way, he found another metal, with chemical properties similar to zinc itself.

We saw earlier how zinc ores and compounds had, in antiquity, often gone by the name *cadmia* when they were produced in copper smelting. Stromeyer took this as his cue for naming the new metal, which he christened *cadmium*. It was not perhaps the most felicitous choice, given the

confusion it invited with the old term for zinc oxide—but there you go. Stromeyer set about studying the chemical behavior of the new element, in the course of which he found that passing hydrogen sulfide gas through a solution of a cadmium compound precipitated a bright yellow solid, cadmium sulfide, which he said "promises to be useful in painting". It surely was, especially when it was discovered how to make an orange form of the compound too. These were sold from the middle of the century as cadmium yellow and cadmium orange—joined, in 1910, by cadmium red. That color contained a small amount of selenium—an element discovered in 1817 by Berzelius, and named after the moon—in place of some of the sulfur. It remains even today probably the artist's favorite and richest red—but is still expensive, and was almost banned in 2014 in Europe because of concerns about cadmium's slight toxicity.

OPPOSITE: Chrome (3–4), arsenic (11–20), and other ores. From Johann Gottlob von Kurr's *The Mineral Kingdom*, Edinburgh: Edmonston and Douglas, 1859, Plate XXII, Science History Institute, Philadelphia.

THE RARE EARTH ELEMENTS

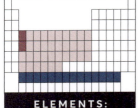

ELEMENTS:

Scandium 21

Yttrium 39

Lanthanum 57

Cerium 58

Praseodymium 59

Neodymium 60

Promethium 61

Samarium 62

Europium 63

Gadolinium 64

Terbium 65

Dysprosium 66

Holmium 67

Erbium 68

Thulium 69

Ytterbium 70

Many elements are named after the places they were discovered, often for reasons of national pride: germanium, francium, polonium. But no place in the world has been a richer source of element names than the small Swedish village of Ytterby, which has no fewer than four elements named in its honor.

Ytterby became a mining village in the late eighteenth century, although quartz had been mined in the region for at least three centuries previously. It was a source of feldspar, an aluminosilicate mineral used in glassmaking and ceramics. In 1787 an army officer and amateur chemist named Carl Axel Arrhenius reported a heavy black mineral that became known as ytterbite. It was first described the following year by Bengt Reinhold Geijer, a Swedish chemist who had mentored Arrhenius in his studies of gunpowder.

Especially dense rocks like this were, as we saw earlier, often suspected of being metal-bearing, and a sample was sent to the Finnish chemist Johan Gadolin at the University of Åbo for analysis. Gadolin reported in 1794 that it contained a new "earth"—meaning a metal compound, usually an oxide. Three years later, the Swedish chemist Anders Gustaf Ekeberg confirmed that discovery and proposed that the "earth" be called yttria, after Ytterby. (It was he who thereby gave the mineral the name ytterbite, although today it is named for Gadolin: gadolinite.) It wasn't until 1828, however, that pure yttrium metal was first made, by the German Friedrich Wöhler.

A most productive earth

There was more than this, however, in the rare mineral from Ytterby. In 1843 a German chemist named Carl Gustaf Mosander, who was Berzelius's assistant and lived in the same house, showed that the yttria extracted from it was actually a mixture of three different oxides: that of yttrium itself (which was white) and also a yellow and a rose-red one. He found that these latter two contained, respectively, new elements that were named terbium and erbium: both contractions of "Ytterby." In 1878 Jean Charles Galissard de Marignac in Geneva found a fourth "earth" (oxide) in yttria, which demanded yet another permutation of the name of its source: ytterbium.

Two years later, while investigating an yttrium mineral from the Ural mountains named samarskite, Marignac discovered two other elements present in small quantities. One had already been discovered in 1879 by the French chemist Paul Émile Lecoq, who named it after the mineral source: samarium. The other was isolated by Marignac in 1880, and he called it alpha-yttrium; but when Lecoq also made it six years later, he suggested to Marignac that it be named after the man who had first begun to reveal what was now evidently an entire family of new elements: gadolinium.

Lying hidden

Carl Mosander's work in the 1840s seriously complicated the roster of elements. At this time he also investigated the oxide of another new metallic element that his mentor Berzelius had identified. In 1803 Berzelius collaborated with the chemist

BELOW: Lennart Halling's photograph of the Ytterby feldspar mine, Resarö, Sweden, ca. 1910, Technical Museum, Stockholm.

Wilhelm Hisinger, whose family owned a mine at Bastnäs in Sweden, on a heavy mineral that Hisinger had brought him. They found in it a new oxide that they named ceria, after the recently discovered asteroid Ceres (actually a dwarf planet). Martin Klaproth in Germany identified ceria at the same time. Ceria is the oxide of the soft, ductile metal cerium. But Mosander later found that the ceria of Berzelius and Hisinger is not pure: it contained at least two other oxides. One of these

RIGHT: Finnish chemist Johan Gadolin after whom the mineral gadolinite is named. Portrait miniature, 1797–1799, Finnish Heritage Agency.

OPPOSITE: German chemist Carl Gustaf Mosander who was pivotal to the discovery of many rare earth elements. Pencil drawing by Maria Röhl, 1842, Royal Library of Sweden, Stockholm.

Mosander named lanthana, after the Greek for "to lie hidden," for it often seemed to accompany cerium in the minerals in which that element was present. The element it contains is lanthanum, which was not made in pure form until the early twentieth century.

The other oxide in ceria was named didymia, meaning "twin," by Mosander. This proved to be a mixture: in part samarium, but also containing two other elements that were separated by Carl Auer von Welsbach in 1885 and named neodymium (new didymium) and praseodymium (green didymium).

A new family

Whence this absurd proliferation of new metallic elements? They, and others—seventeen in all—are called the rare earth elements, and include also scandium, dysprosium, europium, promethium, thulium, holmium, lutetium (the last two are named after their cities of discovery: Stockholm and Paris, which is Lutetia in Latin). Fifteen of these elements fall consecutively in the periodic table, from lanthanum to lutetium, and are known as the lanthanide elements. They tend to occur together in nature because they have very similar chemical properties, forming the same kinds of compounds. That, in turn, is because they all share a similar configuration of their electrons. The lanthanides have a "shell" of electrons with space for 14 in total, which steadily fills up as the series progresses from lanthanum to lutetium—but because this shell is in some sense buried beneath the outermost electron shell, this doesn't much change how the elements behave in chemical reactions. The existence of the lanthanides is, you might say, a quirk of the periodic table—that is, of the rules that govern how electrons are arranged in atoms—which permits a seemingly redundant abundance of elements that are barely distinguishable for most purposes. If that seems profligate of nature—well, the elements were not constructed to flatter our preconceptions about how the world should be constituted.

JOHN DALTON'S ATOMS

In 1887 the English chemist Henry Enfield Roscoe declared that "Atoms are round bits of wood invented by Mr Dalton." He was being gently ironic. Many scientists (although not all of them) believed that atoms were really the fundamental, indivisible particles from which all matter is made. But if so, they were far, far too small to be seen—and so the only images anyone had of an atom were the wooden balls used by the architect of the modern atomic theory, English chemist John Dalton, to illustrate his ideas.

LEFT: English chemist John Dalton. Etching by J. Stephenson, nineteenth century, Wellcome Collection, London.

John Dalton (1766–1844) was a modest school-teacher, educated in the village schools of England's Lake District, who couldn't have gone to the great universities of Oxford or Cambridge even if he'd wanted to, because his Quaker faith excluded him as a religious dissenter. He announced his atomic theory in papers presented to the Manchester Literary and Philosophical Society, of which he was the secretary, between 1803 and 1805. You should publish them in a book, his colleagues advised him—for "the interests of science, and [your] own reputation." And so he did: published in 1808, it had the ambitious title of *A New System of Chemical Philosophy*.

Dalton's theory is often said to have brought up to date an old idea about atoms as building blocks of nature that harks back to the ancient Greek philosophers Leucippus and Democritus. That's true in a way, but old ideas about atoms didn't really explain anything about chemistry. Dalton proposed them to account for why elements seemed often to combine with one another in fixed proportions: they couldn't just be blended like paints. And those proportions were often simple: to make water, say,

a given volume of oxygen gas combined with exactly twice that volume of hydrogen.

Dalton suggests that this made sense if the constituent atoms of elements join into "compound atoms" with simple ratios: partnering one on one, or perhaps one with two. Crucially, Dalton's papers and book included drawings of what these unions look like, with the atoms shown as circles or balls. A "compound atom" (what today we call a molecule) of water, he suggested, is a pairing of an atom of hydrogen with one of oxygen, while an atom of ammonia is a one-to-one union of hydrogen and nitrogen. In public lectures, Dalton's wooden balls served instead as visual aids.

These proportions are wrong, although there was no way of working that out without knowing the relative weights of the atoms. Little by little, these were corrected over the course of the century—water molecules, say, being recognized as having two hydrogen atoms attached to each oxygen.

The New System was not really a new theory of chemistry. For one thing, it couldn't explain why atoms join together in the first place. Henry Roscoe

ABOVE: Five wooden balls used by John Dalton to demonstrate his atomic theory (ca. 1810–1842). Made by Peter Ewart of Manchester, ca. 1810, Science & Industry Museum, Manchester.

rightly said that the significance of Dalton's theory was not so much that he postulated atoms as the indivisible units of matter, but that he proposed that each type of atom has a unique mass. This supplied a distinction between one element and the next—a distinction today enshrined in the concept of atomic number.

Yet it's the atoms that stick in the mind—because in Dalton's works we could see them. They were wooden balls. Dalton himself regarded these pictures and objects purely as teaching tools; he wasn't making any claims about what molecules really look like. Yet "molecular models"—colored plastic balls connected by sticks—are now routinely used to show the actual shapes of molecules.

It wasn't until a hundred years after Dalton's New System was published that scientists found hard evidence that atoms are real. But today we can truly "see" them as blobs in a special type of microscope—and can even use the minuscule needle probe of those devices to shunt and drag atoms around and arrange them into patterns of our own design. They are the balls from which we and everything else are built.

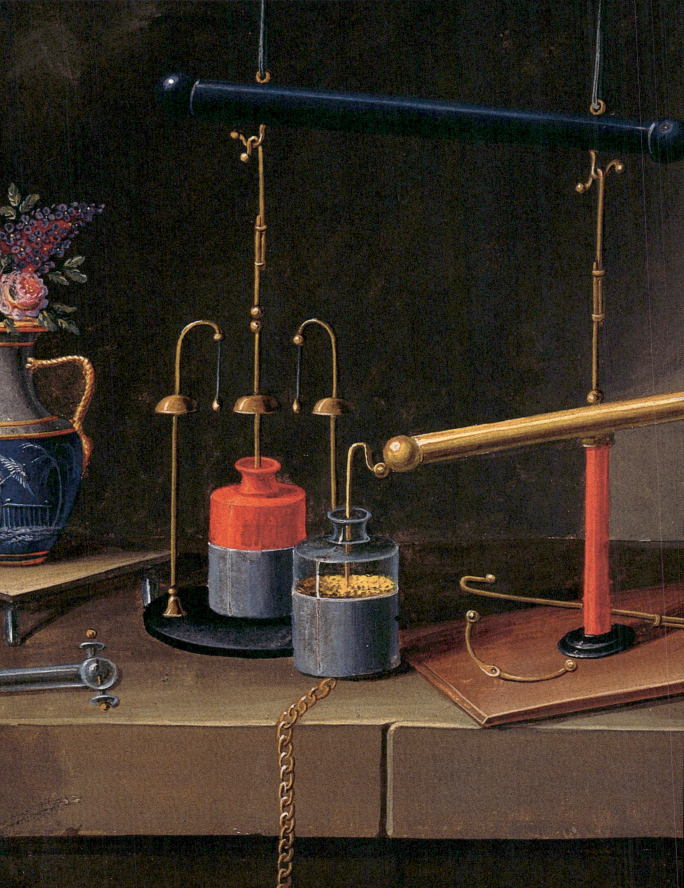

CHAPTER SIX

ELECTRICAL DISCOVERIES

LEFT: Paul Lelong's gouache *Electricity:*
Condenser Jars, an Electro-static Generator,
and a Vase with Flowers, 1820, Wellcome
Collection, London.

ELECTRICAL DISCOVERIES

As the eighteenth century progressed, many scientists increasingly suspected that electricity held deep and profound mysteries. In the early part of that century, the Englishman Stephen Gray showed that static electricity produced by rubbing a glass tube would flow like a fluid along metal wires. He electrified a schoolboy suspended on a platform from the ceiling, and used a metal rod to draw sparks from the poor lad's nose. Spectacular demonstrations like this became a popular trick in the well-to-do parlors and salons of Europe.

Electricity seemed to be some kind of fluid. In 1745 the Dutch scientist Pieter van Musschenbroek, working in Leiden, showed that it could be "collected" by using static electricity, generated by hand-turning a glass sphere on an axle, to charge up metal foil electrodes attached to the inside and outside of a glass jar half-filled with water—a so-called "Leyden jar," which offered a convenient way to store electricity for experiments.

During the 1740s and 1750s, the American scientist and statesman Benjamin Franklin—famed for his kite-flying experiments during thunderstorms (which he probably never actually performed)—made careful studies of how electricity could accumulate in and discharge from Leyden jars. Some of these devices held enough charge to deliver nasty, even life-threatening shocks to the unwary experimenter. In 1767 Joseph Priestley, the English chemist and political reformer who may or may not warrant being considered the discoverer of oxygen, published an immensely popular book summarizing the state of knowledge about electricity.

About two decades later, electricity stored in Leyden jars was discharged through dissected frogs' legs by the Italian physician Luigi Galvani, who observed

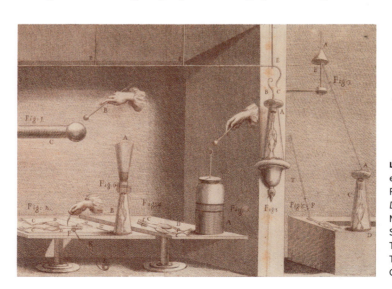

LEFT: Electrical experiments on frogs. From Luigi Galvani's *De Viribus Electricitatis*, Mutinae: Apud Societatem Typographicam, 1792, Table I, Wellcome Collection, London.

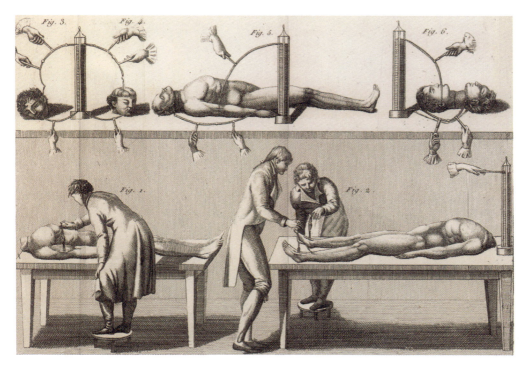

LEFT: Electrical experiments on body parts (by Luigi Galvani's nephew), an inspiration for Mary Shelley's *Frankenstein*. From Jean (Giovanni) Aldini's *Essai Théorique et Expérimental Sur Le Galvanisme*, Bologna: Joseph Lucchesini, 1804, Wellcome Collection, London.

that they twitched in response as if reanimated. He wondered if electricity might in fact be the activating principle of life, a theory that became known as galvanism and which appealed to Mary Shelley when in 1816 she speculated in her nascent novel *Frankenstein* about how a dead body might be brought back to life.

Galvani found that he could also make animal muscles twitch by connecting them to two different metals, such as copper and zinc, placed in contact with one another. In 1800 his compatriot Alessandro Volta at Pavia experimented with stacks of these metals interleaved with salt-soaked pieces of cloth or card (which conduct electricity), and found that this "pile"—really a primitive kind of battery—could produce a substantial and sustained current. While Galvani thought that the metals in his experiments were receiving electricity from the animal tissues, Volta insisted that it worked the other way around: the two metals were the source of the current.

While the two men engaged in a lively and sometimes heated argument about the matter, Galvani's nephew Giovanni Aldini used Volta's pile to conduct some gruesome and literally shocking experiments to "reanimate" much more than dissected frogs' legs. First he used electrical discharges to produce muscle movements—a semblance of life—in bulls' heads brought fresh from the slaughterhouse. And in 1803 he connected his batteries to the corpse of a criminal brought from the gallows at Newgate in London.

Others were finding less controversial and ultimately much more useful applications of the voltaic pile. In 1800 two English scientists, William Nicholson and Anthony Carlisle, studied the transmission of electricity through water and observed gases bubbling from the immersed electrodes. They found that these were oxygen and hydrogen, which Antoine Lavoisier had declared to be the constituents of water itself. The two researchers had used electricity to split water into its elements, a process that was soon named electrolysis.

In other words, electricity could be used to conduct a chemical reaction. Might other substances also be decomposed in this way into their elemental components?

POTASSIUM

GROUP 1

19

K

Potassium

Alkali metal

ATOMIC NUMBER
19

ATOMIC WEIGHT
39.099

PHASE AT STP
Solid

The voltaic pile immediately caught the imagination of a young scientist in England named Humphry Davy. Born to a family of modest means in Cornwall, Davy came to science as something of an outsider, lacking any formal training and being largely self-taught. He became the apprentice of a surgeon in Penzance as a teenager, and then joined the Bristol-based Pneumatic Institute of physician Thomas Beddoes in 1798, who researched the medical effects of gases such as nitrous oxide (laughing gas). There Davy heard about Volta's pile, and constructing one of these batteries for himself, he repeated some of the Italian's experiments.

In 1801 the ambitious Davy left Bristol to interview for the post of lecturer at the newly established Royal Institution in London, where he hoped to continue his studies of "galvanism." He was appointed, and his first lecture in April was on that subject. Davy's public lectures were flamboyant and spectacular, including startling and entertaining practical demonstrations such as the effects of laughing gas. They were so popular—aided no doubt by the youthful Davy's dashing good looks—that crowds flocked to hear them in Albemarle Street, which became London's first one-way street to cope with the busy traffic of carriages.

Mindful of the electrical water-splitting of Carlisle and Nicholson, Davy began to investigate the effects of passing an electrical current from a voltaic pile through chemical solutions and molten salts. He soon found that some metals, such as iron, zinc, and tin, could be extracted from solutions of their salts, appearing as a coating on the negative electrode. But when he tried this with the

RIGHT: Portrait (by an unknown artist) of Sir Humphry Davy, who was knighted in 1812, Wellcome Collection, London.

alkali potash—in today's terminology, potassium hydroxide—he would get only hydrogen at that electrode, as one did from water. In 1807 he tried a different tactic: melting potash and then electrolyzing it with a large, powerful voltaic pile containing a stack of no fewer than 274 plates of copper and zinc. At the positive electrode, oxygen gas bubbled out. But at the negative electrode there appeared "small globules having a high metallic lustre," looking like mercury, "some of which burnt with explosion and bright flame." His cousin, assisting him with the experiment, reported that Davy "bounded about the room in ecstatic delight" at this sight.

Davy found that small pieces of this metal, when collected, would ignite if tossed into water, with a lilac-purple flame, while dashing around on the surface. With larger pieces there would be "an instantaneous explosion…with brilliant flame"—and all that remained was a solution of potash. He revealed this dramatic process before a rapt audience for a prestigious lecture he gave that year at the Royal Society.

Davy concluded that this inflammable metal was a fundamental ingredient of potash—a new element that he called potassium. It is set alight by the reaction of potassium with moisture in the air, a process that generates hydrogen gas—which is then ignited by the heat of the reaction. Because it came from an alkali (and reverts quickly to the alkaline oxide or hydroxide in air and water), it was called an alkali metal.

A German translation of Davy's report based the name on the more common German term for potash: *kali*, giving instead *Kalium*. When he tried to standardize chemical nomenclature in 1811, Jöns Jacob Berzelius preferred the German form, and so he gave potassium the rather confusing symbol K.

RIGHT: An early voltaic pile, possibly made by Alessandro Volta, nineteenth century, Science Museum, London.

SODIUM

Days after discovering potassium, Humphry Davy tried electrolysis on another alkali, molten sodium hydroxide (known then as soda). Again he saw the formation of a highly reactive metal at the negative electrode, which he called sodium. "When thrown upon water," Davy reported, "it effervesces violently, but does not inflame."

The alkali substances potash and soda had been among the chemist's most useful substances for millennia—although those terms had traditionally referred not to the hydroxides but the carbonates of the respective metals. Mixed with sand, they were used in glassmaking, where they lower the temperature that is needed to melt the quartz of the sand. Boiled with animal fats, they can be used to produce soap.

Potash was, as the name implies, typically made from ashes of wood or other plants. (*Al-kali* is itself an Arabic word meaning "ash"; from the root term *kali* comes the name and symbol Berzelius gave to potassium.) Most types of plant produce potassium-rich ashes, but some have more sodium.

¶Sera. Eſt abſterſiu⁹ bumoꝛ groſſoꝛ. ꝛ ab
ſtergit ꝛ lauat ꝛ ꝓfert caſui vulue ꝛ ſquinãtie.
¶Et diminuit albedinẽ oculi ſeu pannũ ei⁹.

Soda, however, could also be found in mineral form—the Greeks called it *natron* or *nitron*, from which, confusingly, the term niter is derived. This is the origin of the chemical symbol of sodium (Na), because the German chemists likewise preferred to call Davy's new element *natronium* or *natrium*.

The earliest natron/nitron/niter, collected from the Natron Valley in Egypt or by evaporation of waters from the Nile in sun-baked pits, was a mixture of sodium carbonate with common salt (sodium chloride). It was used for washing and, along with lime (calcium carbonate), for glassmaking. Pliny the Elder in the first century AD asserts that the production of glass was discovered by accident when some merchants who were trading in natron used blocks of it to support their pans on a cooking fire on a sandy shore and found that a clear liquid ran out from the ashes and set

hard. This is probably as fanciful as many of Pliny's stories, but the discovery of glass was probably accidental in some manner.

Right up until the late eighteenth century, alkalis were still classed according to their source: vegetable alkali from ashes, mineral alkali from rocks like natron (or natrum, as it was sometimes then called). The French chemists in Antoine Lavoisier's circle promoted the use of "soda" (known as *soude* in French) over "natron," saying that it "was more universally known." If soda or potash was mixed with "slaked lime" (calcium hydroxide), they became even more corrosive alkalis: the hydroxides of sodium and potassium (as we would say now), known sometimes even today as caustic soda and caustic potash. In fact, *soude* or *potasse* often referred to these hydroxides themselves. It was clear that sodium and potassium carbonate weren't themselves elemental substances, because one could extract carbon dioxide ("fixed air") from them.

Yet was caustic soda itself an element? The French chemists suspected not, and figured that new elements might be forthcoming if a way could be found to break it down. It was this expectation that inspired Humphry Davy to see what his voltaic pile might summon forth from the molten substances, and which led him to isolate the two alkali metals in 1807.

RIGHT: Engraving showing Sir Humphry Davy using electrical decomposition to discover potassium and sodium, ca. 1878, World History Archive.

U.PARENT

CALCIUM, MAGNESIUM, BARIUM, AND STRONTIUM

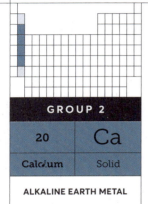

GROUP 2	
20	Ca
Calcium	Solid

ALKALINE EARTH METAL
ATOMIC WEIGHT: 40.078

GROUP 2	
12	Mg
Magnesium	Solid

ALKALINE EARTH METAL
ATOMIC WEIGHT: 24.305

GROUP 2	
56	Ba
Barium	Solid

ALKALINE EARTH METAL
ATOMIC WEIGHT: 137.327

GROUP 2	
38	Sr
Strontium	Solid

ALKALINE EARTH METAL
ATOMIC WEIGHT: 87.62

Another common alkaline material with a long history of practical use was lime, which is calcium carbonate. Lime is abundant in mineral forms: it is the fabric of chalk, limestone, and marble. They all arise from the same source: the skeletal remains of marine organisms like molluscs and single-celled microscopic organisms such as foraminifera and dinoflagelates, which secrete calcium carbonate to make their shells or protective exoskeletons. When these organisms die, their protective biomineral shells sink to the bottom of the ocean and become sedimentary rock. Compressed in the Earth, these sediments are transformed first into chalk, then (at higher pressures) limestone and, finally, dense marble. Bird eggshells are also made from calcium carbonate.

Lime has long been mined. One of its principal uses, when mixed with sand (or fine gravel) and water, was to make the mortars that bound the masonry in buildings. The word lime comes from the Latin *limus*, which means a sticky mud or slime. The limestone would be baked in a kiln, driving off carbon dioxide and leaving behind "quicklime," which is calcium oxide. This was then "slaked": added to water in order to make a slurry, which is composed of calcium hydroxide. When exposed to air, the slaked lime gradually reacts with atmospheric carbon dioxide and reverts to calcium carbonate, setting with mineral-like hardness.

BELOW: A section of Roman concrete consisting of lime, volcanic sand, and rock. From the aqueduct of Fréjus, Mons, first century AD.

ABOVE: Detail of W. H. Payne's aquatint *Men Working a Lime-kiln*, 1804, Wellcome Collection, London.

BELOW: Limestone bricks and lime mortar line the corridors of the world's oldest Step Pyramid, built by Imhotep as a mortuary for the pharaoh Djoser, at Saqqara, Egypt, ca. 2670–2650 BC.

Lime mortar, as well as mortar made from the mineral gypsum (calcium sulfate), was used in the Egyptian pyramids. The Romans made a more durable mortar by adding a special volcanic ash that reacted with slaked lime to form a kind of concrete. The Roman engineer Vitruvius listed recipes for these robust mortars, which have held together some Roman buildings to this day.

The importance of mortar in construction, as well as quicklime's use in making soaps and textiles, meant that the substance (a strong, caustic alkali) was produced on a larger scale in antiquity than any other apart from common salt. Throughout the eighteenth century, chemists puzzled over its causticity, and whether it was related to alkalinity: when quicklime was slaked, it stayed alkaline but was no longer caustic. Joseph Black discovered that slaked lime could supply a test for carbon dioxide (which he called "fixed air"): as the gas bubbles through it, the calcium combines with it to form

OPPOSITE: Carbonate of strontium. From James Sowerby's *British Mineralogy, or, Coloured Figures Intended to Elucidate the Mineralogy of Great Britain*, London: R. Taylor and Co., 1802–1817, Plate LXV, Smithsonian Libraries, Washington, DC.

the insoluble calcium carbonate, turning the liquid a chalky white.

Antoine Lavoisier's list of thirty-three elements in his *Traité Élémentaire* of 1789 included chalk (*chaux*, or calcareous earth) under the heading of "earths." But in an English translation of the book in 1793, a note was added to say that experimenters in Hungary claimed to have extracted a metal from chalk which they proposed to call parthenum. The translator, Robert Kerr, said that a better name, more consistent with the French system of nomenclature, would be calcum. The Hungarian claim was soon disproved by the German chemist Martin Klaproth, who showed that the metal made this way was probably iron. But Humphry Davy shared the suspicion that there was a metal lurking in this "alkaline earth," and in 1808 he used electrolysis to find out.

Davy couldn't check in the same way as he did for potassium and sodium hydroxide—by electrolyzing them when molten—because if you heat calcium hydroxide or carbonate, you just make quicklime (the oxide), which won't melt. Instead, he passed the current of his voltaic pile through a powdered mixture of "calcareous earth" (quicklime) and mercury oxide, moistened with water. This produced a small puddle of liquid mercury at the negative electrode. When Davy collected this and heated it to evaporate the mercury, he found a metallic residue left behind, which had been present in the mercury as an amalgam. Heeding Kerr's advice (more or less), he called it calcium.

Calcareous earth wasn't the only substance Davy investigated. Lavoisier's "earths" also included *magnésie* (magnesia) and *baryte*, two other mild alkalis that could be obtained from minerals. We saw earlier that magnesia was sometimes confused with manganese compounds, both of them derived from Magnesia, the region in Anatolia where such minerals were mined. Davy used his mercury oxide method of electrolysis with these earths too, and again found he could produce a mercury amalgam of

ABOVE: Map detail showing the rich mining district of Strontian, Argyllshire. From Richard Cooper after Alexander Bruce, *A Plan of Loch Sunart…become Famous by the Greatest National Improvement this Age has Produc'd*, Edinburgh: Bruce, 1733, National Library of Scotland, Edinburgh.

new metals. At first he named one of them magnium, noting that "magnesium" had already been misused to refer to manganese. But after "candid criticisms of some philosophical friends," Davy relented in his 1812 book *Elements of Chemical Philosophy*, accepting after all the now familiar name. Davy noted that making the amalgam of magnesium and mercury took longer than it did for the other metals, but he later found he could decompose magnesia more directly by heating it with potassium vapor in a platinum tube and dissolving the residue—a "dark gray metallic film"—in mercury.

Davy included one more earth (which is to say, a metal oxide) in these experiments too: strontia, made from a mineral called strontianite that was identified in 1790 at the lead mine in Strontian, west Scotland. The corresponding metal was named strontium. So in one swoop, Davy discovered an entire family of new metals: calcium, magnesium, barium, and strontium, collectively known (with beryllium, the lightest member of the family) as the alkaline earth metals.

BORON

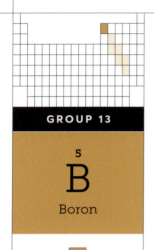

5

B

Boron

Metalloid

ATOMIC NUMBER
5

ATOMIC WEIGHT
10.81

PHASE AT STP
Solid

Among the historically important chemical commodities such as saltpeter, quicklime, and potash was one known as borax, a white salt that could be found in mineral form. It was mentioned by some Arabic alchemists from around the eighth century (the word comes from the Arabic term *buraq*, meaning "white"), and there were natural deposits of the substance in central Asia, although most borax was imported from Tibet along the Silk Road. It was used in goldmaking (it acted as a "flux" that helped the metal to melt), glassmaking, and as a medicine. But borax was easily confused with other white salts, for no one knew what was in it. In the early eighteenth century, the French chemist Louis Lémery pronounced it the least understood of all the naturally occurring salts.

Antoine Lavoisier's list of elements included boron as a "boracic radical," meaning that he considered it a component of an acid (boracic acid, which today we call boric acid), then used as a sedative agent. Borax and its related compounds were known to burn with a green flame, which is how deposits of the mineral were identified in Italy and the United States in the eighteenth and nineteenth centuries.

Borax was an obvious candidate for Humphry Davy's quest for new elements. In October 1807, he electrolyzed "slightly moistened" boric acid and saw a "dark olive coloured" substance form at the negative electrode. He called this element boracium, and presumed that it was a metal—but changed the name to boron once his subsequent studies showed this to be mistaken, the -ium suffix being used exclusively for metals. Boron, Davy said, "is more analogous to carbon than to any other substance."

Electrolysis only produced tiny amounts of boron, but the following March Davy found another way to produce it more abundantly: by heating boric acid with potassium metal in an iron or copper tube.

If Davy was the first to make this element, he wasn't the first to report it. His successes in identifying sodium and potassium brought him acclaim even in England's military enemy France, where Napoleon Bonaparte awarded him with a prestigious prize. But Napoleon was also keen to see French scientists making such discoveries, to which end the French leader supplied a large voltaic pile to Louis-Joseph Gay-Lussac and Louis-Jacques Thénard in Paris. They too began to investigate substances like borax, but could extract nothing of interest. Nine days before Davy presented his results on June 30, 1808, however, Gay-Lussac and Thénard reported a new element obtained by the same means as Davy: heating boric acid with potassium. They called it *bore*, the name by which boron is still known in French.

The truth is that neither party had actually isolated pure boron: their samples probably contained around 50 percent of other elements. A sample of nearly pure boron was not made until 1892, when Henri Moissan made it

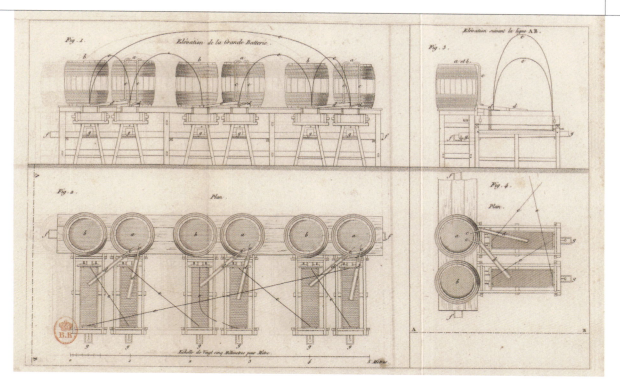

by reacting boron oxide with magnesium metal. A researcher at the General Electric Company in the United States named Ezekiel Weintraub made a still purer form of boron by passing sparks through a vapor of boron trichloride and hydrogen in 1911. But truly pure boron was not made until the late 1950s.

As Davy eventually realized, boron is a nonmetal: it does not conduct electricity, and has a dull, dark gray appearance. To be frank, some chemists regard boron as warranting its name by virtue of it being a supremely *boring* element. But that's unfair: pure boron can adopt an unusually wide range of different crystal structures, some of them very complicated and based on clusters of twelve boron atoms joined into the shape of polyhedral clusters of twelve boron atoms. Boron is also a component of some of the hardest substances known: boron carbide and boron nitride. The former is used in tank armor and bulletproof vests; the latter (known industrially as borazon) is second only in hardness to diamond and therefore of great value in cutting and abrasion tools.

RIGHT: An undated photograph of Ezekiel Weintraub, a researcher at the General Electric Company, Williams Haynes Portrait Collection, Box 16, Science History Institute, Philadelphia.

ABOVE: Distillation equipment used by Louis-Joseph Gay-Lussac and Louis-Jacques Thénard in their preparation of boron. From their *Recherches Physico-Chimiques*, Paris: Chez Deterville, 1811, Plate 2, National Library of France, Paris.

ALUMINUM, SILICON, AND ZIRCONIUM

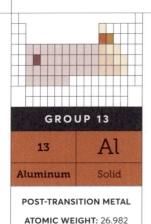

GROUP 13

13	Al
Aluminum	Solid

POST-TRANSITION METAL

ATOMIC WEIGHT: 26.982

GROUP 14

14	Si
Silicon	Solid

NON-METAL

ATOMIC WEIGHT: 28.085

GROUP 4

40	Zr
Zirconium	Solid

TRANSITION METAL

ATOMIC WEIGHT: 91.224

Humphry Davy also turned his attention to two other long-known mineral substances that were suspected of being compounds harboring unknown elements. They were called alumina and silica—or as Davy called them, alumine and silex. They were both among the "earths" long recognized by chemists. Alumina was associated with the ancient salt alum used in dyeing and tanning; silex was named after the Latin for "flint," and seemed to be the fabric of sand.

Yet when Davy tried to decompose these materials into their elements by melting them and using electrolysis, he got nowhere. He was, he wrote, "obliged to seek for other means of acting upon them." He mixed alumina and potash in a platinum crucible and electrolyzed them together, reporting that this produced "a film of a metallic substance" on one of the platinum electrodes, which decomposed in acid to reconstitute alumina.

Then he tried the reaction by which he had obtained boron from boric acid: heating both silex and alumina with potassium vapor. In the former case, he produced "a grayish opaque mass, not possessed of metallic splendour," and "black particles not unlike plumbago [graphite]"; in the latter, he saw "numerous gray particles, having the metallic lustre."

Davy was cautious enough not to jump to conclusions. He suspected that in both cases he had seen signs of new elements, but he knew that he would need to be able to isolate them and fully investigate their chemical behavior before he could be sure (and could convince others). All the same, he tentatively proposed names for these new elements: alumium and silicium. He also carried out the same experiments with the mineral known as zircone, and saw hints (again, no more than that) of what he believed to be another new metal, which he called zirconium.

RIGHT: C. A. Jensen's *Portrait of the Scientist Hans Christian Ørsted*, 1832–1833, Statens Museum for Kunst, Copenhagen.

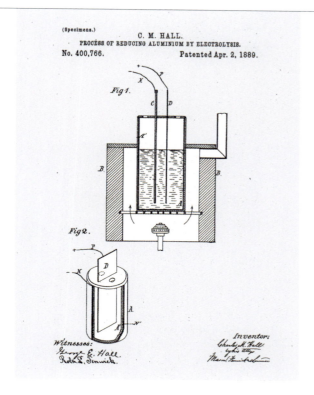

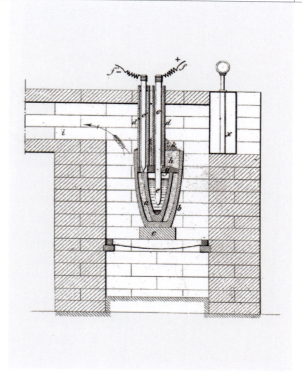

Davy was wise to be careful with his words, because whatever those gray and black particles were, they do not seem to have been the pure elements Davy was hoping for. Gay-Lussac and Thénard tried their trick of reacting a compound derived from silex with potassium metal in 1811, but they don't seem to have made anything more than a very impure form of silicon either. Not until 1823 was relatively pure silicon first made, when the Swedish chemist Jöns Jacob Berzelius heated silicon fluoride with potassium and produced a gray powder that he identified with Davy's putative silicium. Pure aluminum was probably first made in 1835 by the Danish scientist Hans Christian Oersted by reacting aluminum trichloride with potassium vapor, a process perfected by the German Friedrich Wöhler using sodium instead of potassium.

Davy himself modified the name alumium to aluminium (or, alternatively, aluminum, which remains the preferred spelling in the United States).

His silicium did not turn out to be a metal, as he'd anticipated, but has the important property of semiconductivity: it conducts electricity only very slightly, and to a degree that increases a little when it is warmed. Already in 1817, before Berzelius's experiment, the Scottish chemist Thomas Thomson had pointed out that "there is not the smallest evidence for [the] metallic nature" of silicium, and so proposed that it be instead named by analogy with carbon and boron: silicon.

The semiconducting nature of silicon is what makes it so valuable for modern electronics. Metals conduct indiscriminately: a current passes through them freely. But the passage of a current through silicon can be precisely controlled, for example by adding impurities (dopants) to silicon's crystal lattice, or using electric fields. This means that currents can be switched on and off, making silicon just the material for electronic switches called transistors that form that basis of almost all

microelectronic circuitry today. These devices can be etched onto slabs (chips) of crystalline silicon at the microscopic scale, so that chips the size of a thumbnail can now carry many millions of transistors. The steady shrinkage of silicon transistors, allowing them to be packed ever more densely onto chips, underpins the explosion in the processing power of computers and handheld electronics.

Silicon is made industrially by reducing silica—molten sand—with charcoal, much like the smelting of many metals. But making silicon with the extreme purity needed for microelectronics is another matter, and relies on a technique called zone refining in which the impurities in raw silicon are slowly sequestered into a molten zone that passes along a silicon rod.

Aluminum, meanwhile, has proved its worth as the lightest of all abundant, strong metals, making it an ideal structural material. Both silicon and aluminum are found in many rocks and minerals, where they combine with oxygen atoms to form crystalline networks of strong chemical bonds: aluminosilicates. So they are available in almost limitless quantities in principle. But extracting them is hard and energy-intensive, so readily and strongly do they combine with oxygen. The main ore of aluminum is the mineral bauxite, which is the oxide. The metal is separated by electrolysis, but bauxite itself has a very high melting temperature of over 3,632°F, and so it must be mixed with an aluminum salt called cryolite to lower this temperature. The process was devised in 1886 in another of those instances of simultaneity that seem to recur in element discovery: both Charles Hall, an American experimenter in Ohio, and Paul-Louis-Toussaint Héroult in France, filed for patents on the method within weeks of one another. After a legal tussle, Hall acquired the US rights, Héroult the European ones.

RIGHT: The loading of aluminum ore at a refining plant for bauxite mining, in Arkansas, 1908, Bettmann Archive.

THE PERIODIC TABLE

We search for order, for systems and classifications to bring some structure to the profusion of the world. That was surely the impulse behind the ancient notion that there were just four fundamental elements (or perhaps fewer still…) from which all else was made. But as the list of elements began to grow, new organizing principles were needed. Antoine Lavoisier's list of elements in 1789 made an ad hoc division into gases and fluids, metals, nonmetals, and earths—but there was no obvious pattern to it all.

LEFT: Undated photograph of Dmitri Mendeleev, in middle age. From the Edgar Fahs Smith Collection, Kislak Center for Special Collections, Rare Books and Manuscripts, University of Pennsylvania.

In 1829, however, the German chemist Johann Wolfgang Döbereiner thought he could glimpse one. Some elements seemed to cluster into triads with similar properties: the alkali metals lithium, sodium, and potassium, say, or the pungent halogens chlorine, bromine, and iodine. A chemistry textbook written in 1843 by Leopold Gmelin in Heidelberg listed ten such triads, as well as other groups of four and five. And in the 1850s, English chemist William Odling listed several groups of elements that shared affinities, such as nitrogen, phosphorus, arsenic, antimony, and bismuth. Elements seemed to come in families.

There was, at the same time, a natural way to order the elements sequentially: by their atomic weights, meaning the weights of equal numbers of atoms of each element. Chemists in the nineteenth century couldn't count atoms, but they relied on the proposal of the Italian scientist Amedeo Avogadro that equal volumes of gas at the same pressure and temperature contained equal numbers of atoms or molecules. Hydrogen was the lightest element, and nearly all the other elements had atomic weights that seemed close to a whole-number multiple of that of hydrogen: carbon is 12 times more weighty, oxygen 16, sulfur 32, and so on. This led the Englishman William Prout in 1815 to suggest that hydrogen was a kind of primal substance from which all others are made, like the prote hyle (see page 16) of the ancient Greeks. (He was pretty much correct, as we'll see.)

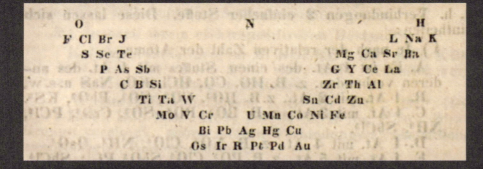

LEFT: Triads (and other groupings) of related elements. From Leopold Gmelin's *Handbuch der Chemie*, Heidelberg: Winter, 1843 Vol. 1, Bavarian State Library, Munich.

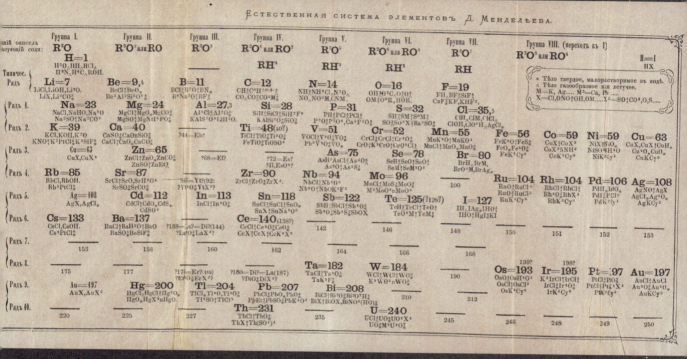

ABOVE: First published periodic table in modern form with each element group arranged vertically. From Dmitri Mendeleev's *Osnovy Khimii* (Principles of Chemistry), St. Petersburg: Tip. t-va Obshchestvennaya polza, 1871, Science History Institute, Philadelphia.

In 1860 another Italian, Stanislao Cannizzaro, unveiled an improved list of atomic weights (relative to hydrogen) at an international conference, based on Avogadro's latest work. When the German chemist Julius Lothar Meyer saw that list, he said: "It was as though the scales fell from my eyes." He realized how this sequential ordering could be combined with the grouping of elements into families, by placing them in a *table*. The weights steadily increased from left to right and from row to row, while the families appeared as vertical columns. He presented this scheme in his 1864 textbook *The Modern Theory of Chemistry*.

Meyer was not alone. Odling suggested a rather similar scheme in that same year, while the English chemist John Newlands pointed out that the sequential list of elements by atomic weight seemed *periodic*: elements shared properties with those eight or sixteen places further on. But when Newlands presented his idea in 1866 and drew an analogy with the octave organization of the musical scale, it was mocked as far-fetched.

There is little doubt, then, that what became known as the Periodic Table of the elements was a fairly well established, if controversial, idea some years before the date of its "official" discovery in 1869 by the Russian chemist Dmitri Mendeleev (1834–1907), who was working at the University of St. Petersburg.

Mendeleev's breakthrough has the advantage today of having a good story attached to it. A Siberian from remote Tobolsk with the wild hair and beard of a hermit, Mendeleev is said to have sought order from Avogadro's improved atomic weights by writing the elements on cards and arranging them like a solitaire game. Exhausted by his unsatisfactory results, he fell asleep in his study on February 17, 1869.

	4 werthig	3 werthig	2 werthig	1 werthig	1 werthig	2 werthig
	—	—	—	—	Li $= 7{,}03$	(Be $= 9{,}3$?)
Differenz $=$	—	—	—	—	$16{,}02$	$(14{,}7)$
	C $= 12{,}0$	N $= 14{,}04$	O $= 16{,}00$	Fl $= 19{,}0$	Na $= 23{,}05$	Mg $= 24{,}0$
Differenz $=$	$16{,}5$	$16{,}96$	$16{,}07$	$16{,}46$	$16{,}08$	$16{,}0$
	Si $= 28{,}5$	P $= 31{,}0$	S $= 32{,}07$	Cl $= 35{,}46$	K $= 39{,}13$	Ca $= 40{,}0$
Differenz $=$	$\frac{89{,}1}{2} = 44{,}55$	$44{,}0$	$46{,}7$	$44{,}51$	$46{,}3$	$47{,}6$
	—	As $= 75{,}0$	Se $= 78{,}8$	Br $= 79{,}97$	Rb $= 85{,}4$	Sr $= 87{,}6$
Differenz $=$	$\frac{89{,}1}{2} = 44{,}55$	$45{,}6$	$49{,}5$	$46{,}8$	$47{,}6$	$49{,}5$
	Sn $= 117{,}6$	Sb $= 120{,}6$	Te $= 128{,}3$	J $= 126{,}8$	Cs $= 133{,}0$	Ba $= 137{,}1$
Differenz $=$	$89{,}4 = 2.44{,}7$	$87{,}4 = 2.43{,}7$	—	—	$(71 = 2.35{,}5)$	—
	Pb $= 207{,}0$	Bi $= 208{,}0$	—	—	$(Tl = 204$?)	—

ABOVE: Periodic table. From Julius Lothar Meyer's *Die Modernen Theorien der Chemie*, Breslau: Maruschke & Berendt, 1864, Wellcome Collection, London.

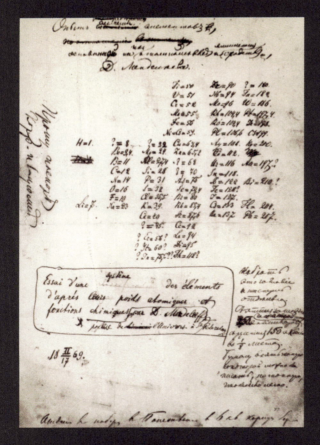

LEFT: Manuscript of Dmitri Mendeleev's first periodic system of elements, February 17, 1869, Science Museum Library, London.

"I saw in a dream a table where all the elements fell into place as required," he was later reported to have said. On waking, he hastily wrote down his vision, and two weeks later he published his "Suggested System of the Elements." But historians of science are skeptical that it really happened this way—for one thing, the "dream" account came from a colleague of Mendeleev's forty years after the event. He would surely have already known of the ideas others had voiced about element families.

Yet the scheme wasn't perfect. To make it fit, Mendeleev needed to take some liberties to ensure that elements with similar chemical behavior fell into the same group—for example, by asserting that some of the accepted formulae for chemical compounds (that is, the ratios in which the elements combined) were wrong. That isn't to suggest he cheated; on the contrary, Mendeleev shows that sometimes in science it's worth sustaining a good

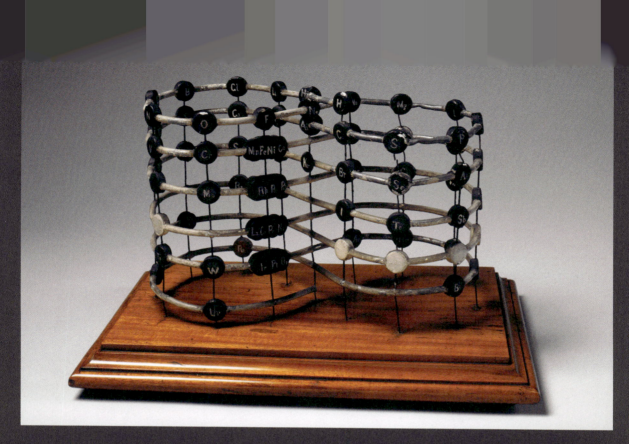

ABOVE: Sir William Crookes's spiral model for illustrating the periodic table, 1888, Science Museum, London.

idea even if it doesn't quite seem to fit with experimental evidence.

Although Lothar Meyer had already drawn up more or less the same periodic table in 1868, he didn't publish it until 1870—and so Mendeleev tends to get the credit, despite Meyer's protestations about his priority. But Mendeleev's recognition doesn't just rest on that lucky timing. He also had the acumen to realize that, for his ordering scheme to work, he had to leave some gaps in the table: in effect, a prediction that other elements existed that were yet to be discovered. (In fairness, Meyer left gaps too, but didn't really establish them as predictions.) It was only when these predictions began to be borne out that Mendeleev's periodic table started to attract wider attention.

The first of these anticipated elements to be found was gallium, discovered in 1875 by French chemist Paul-Émile Lecoq. With relative atomic weight 68, it fitted perfectly into the space Mendeleev had left below aluminum, for which he'd assigned the provisional name "eka-aluminum." Another of Mendeleev's predicted elements, labeled eka-silicon, was discovered in

1886 and christened germanium.

As the periodic table was filled out, it became clear that its periodicity was rather complicated. The first two main rows, from lithium to chlorine, fitted the eightfold pattern well enough, but after that the scheme was interrupted by the "transition metals" such as iron, nickel, and copper. Why the elements fit this scheme at all was a mystery until the internal structure of atoms—the existence of their subatomic constituents electrons, protons, and neutrons—was elucidated in the early twentieth century. Chemical periodicity arises from the way the electrons— which are responsible for the chemical properties— are arranged into shells, a fact only explained when the theory of quantum mechanics was developed between the 1900s and the 1930s. In this way, the periodic table encodes the deepest principles of how atoms themselves are constituted.

OVERLEAF: The periodic table gets creative: Edgar Longman's mural from the 1951 Festival of Britain Science Exhibition (restored by Philip Stewart in 2004).

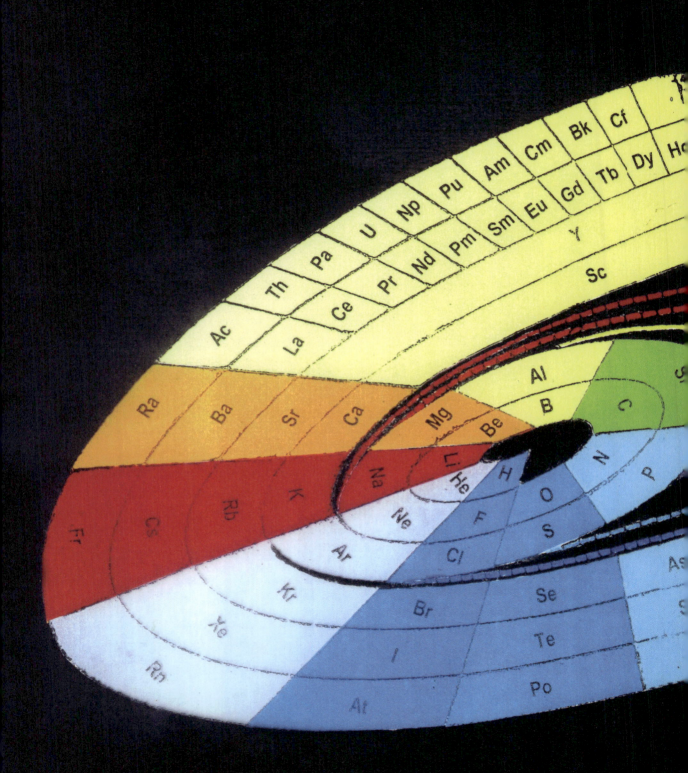

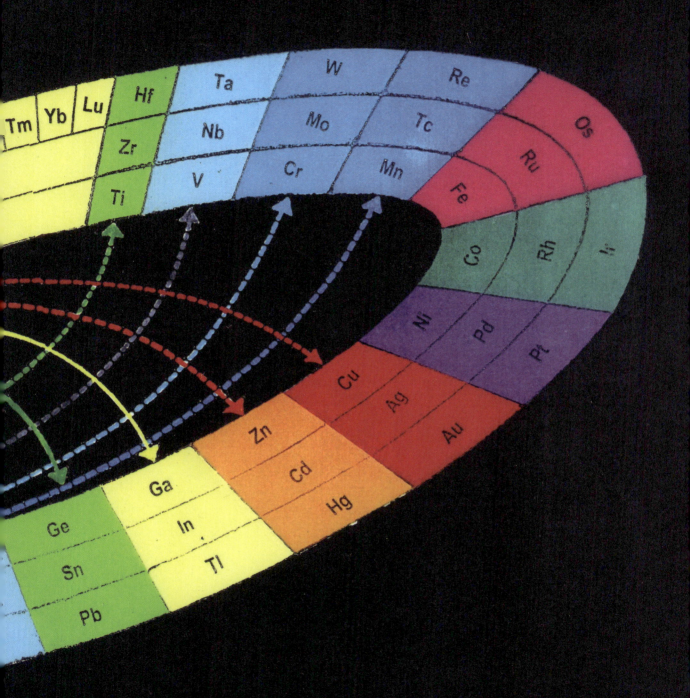

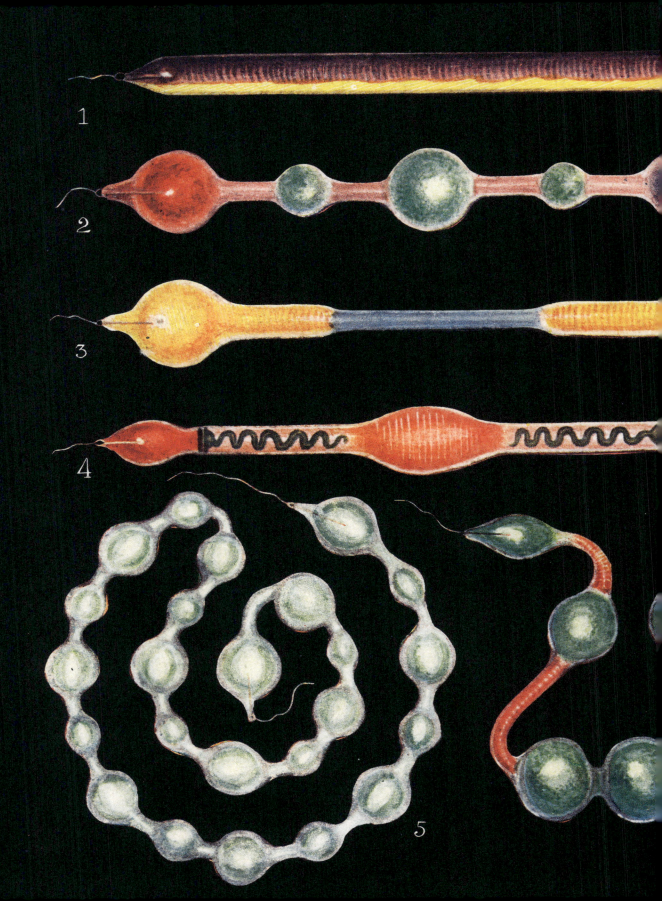

CHAPTER SEVEN

THE RADIANT AGE

THE RADIANT AGE

Aristotle's heavenly fifth element, the aether, had never quite vanished from natural philosophy, and in the mid-nineteenth century it flourished again in a new guise: as the bearer of light, the "luminiferous ether."

Light had always been a contested affair. For Isaac Newton in the late seventeenth century it was composed of a stream of particles, whereas his rival Robert Hooke insisted that it was a wave: "nothing," Hooke wrote in 1672, "but a pulse or motion propagated through an homogeneous, uniform, and transparent medium." Antoine Lavoisier included light in his list of elements in 1789, without specifying its nature. But Hooke's view prevailed; in the early 1800s the English polymath Thomas Young demonstrated how light rays passing through two narrowly spaced slits interfere with one another to cause a pattern of bright and dark bands, a phenomenon that seems to demand a wave picture of light. But waves must have some medium to carry them, as Hooke said—and the ether was resurrected to serve that role. It was thought to pervade the universe, invisible and too tenuous to measure or weigh.

In the 1860s, the Scottish scientist James Clerk Maxwell explained how this worked. He showed that disturbances in electric and magnetic fields travel through space at the speed of light, and supposed that this is, in fact, what light is: an electromagnetic vibration, propagating through the etheric medium that supports these fields, just like sound waves traveling through air. It was this ether that bears light across the vacuum of space. "The vast interplanetary and interstellar regions," Maxwell wrote, "will no longer be regarded as waste places in the universe…We shall find them to be already full of this wonderful medium [which] extends unbroken from star to star."

Maxwell formulated a set of equations to describe these electromagnetic waves. It was immediately clear, however, that the theory didn't impose any limit on the wavelengths and frequencies of the waves. The wavelengths of visible light could be measured, and fell between (in today's units) around 400 billionths of a meter for violet light and 700 billionths for red light.

RIGHT: James Clerk Maxwell, 1881. Engraving by G. J. Stodart after J. Fergus, Wellcome Collection, London.

But the wavelengths of electromagnetic vibrations could, in theory, be longer and shorter than this range. Long-wavelength vibrations (perhaps several meters to several kilometers) were first detected in 1887 by the German physicist Heinrich Hertz, and became known as radio waves. Just nine years later, Italian inventor Guglielmo Marconi showed that the long reach of radio waves could be used for transmitting messages over great distances, broadcast from a source of such waves to a device that could detect them. In 1892 the English chemist William Crookes wrote that radio waves could be used for "telegraphy without wires, posts, cables, or any of our present costly appliances." A few decades earlier, a telegraph cable had been laid, with great expense and difficulty, across the Atlantic seafloor; but now messages could be simply sent—literally, most scientists believed—through the ether. In 1901 Marconi broadcast a radio signal from Cornwall in western England to Newfoundland in Canada.

Electromagnetic waves could also have wavelengths shorter than those of visible light. Ultraviolet light, just beyond the violet end of the spectrum, had been known since 1801, when the German scientist Johann Wilhelm Ritter found that this "invisible" form of light, separated out with a prism, could trigger the darkening of silver salts just as normal light does. (A few decades later this process was used as the basis of photography.) It was disputed for a time whether ultraviolet radiation was of the same nature as ordinary light at all, but Maxwell's theory helped to rationalize it. In 1895 German physicist Wilhelm Roentgen discovered another kind of invisible radiation that could also darken photographic emulsion: he called them X-rays, and they turned out to be electromagnetic vibrations with wavelengths hundreds of times shorter than those of visible light.

By the end of the nineteenth century, it seemed that the world was pervaded with invisible rays, many of them revealed by photography. Some of these "emanations" proved a figment of the imagination, such as the N-rays postulated in 1903 by the French physicist Prosper-René Blondlot; others, like the mysterious "uranic rays" (see page 100) found to come out of uranium salts, or cosmic

ABOVE: Recreation of Guglielmo Marconi's first radio transmitter, built in August 1895. From *Radio Broadcast* magazine, New York: Doubleday, Page & Co., 1926, Vol. 10.

rays streaming from space, presaged important discoveries in early twentieth century science. For some scientists these rays held quasi-miraculous properties—Crookes argued that Maxwell's ether might carry messages not just over the Atlantic, but also between our world and a spirit plane, channeled by another kind of medium: the spiritualists who held séances to communicate with the dead.

In the radiant world of the late nineteenth century, anything seemed possible.

CESIUM AND RUBIDIUM

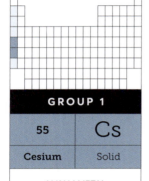

GROUP 1	
55	Cs
Cesium	Solid

ALKALI METAL
ATOMIC WEIGHT: 132.905

GROUP 1	
37	Rb
Rubidium	Solid

ALKALI METAL
ATOMIC WEIGHT: 85.468

During the late Victorian era, a new method for discovering elements was devised which no longer required researchers to separate and purify tangible quantities of new elements. Instead, it depended on the way different elements absorb or emit light of particular colors in narrow bands of wavelength: a method now called spectroscopy.

It was invented by two German scientists, the physicist Gustav Kirchhoff and chemist Robert Bunsen, in 1859. Bunsen knew that different metal elements will emit light of characteristic colors when heated in a flame: a convenient method for figuring out which elements were present in a compound, say. But rather than rely on a visual inspection of the flame's color, Kirchhoff and Bunsen devised an instrument, called the spectroscope, that used a prism to separate this light into its different wavelengths, just as Isaac Newton split natural sunlight into the rainbow spectrum. Sunlight contains all the spectral colors, but the researchers showed that the colored light from

LEFT: Gustav Kirchhoff (left) and Robert Bunsen (seated), with chemist Henry Roscoe in 1862. Photograph by Emery Walker from H. E. Roscoe's *The Life & Experiences of Sir Henry Enfield Roscoe…Written by Himself,* London: Macmillan Company, 1906, Science History Institute, Philadelphia.

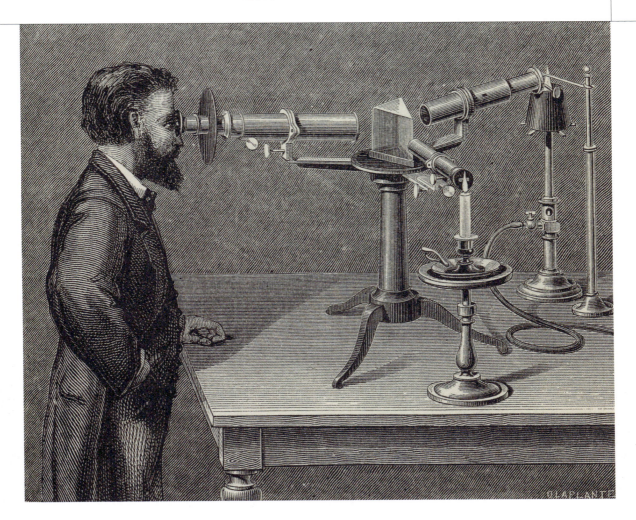

a metal burning in the gas flame of one of Bunsen's iconic burners contains distinctive bright bands at wavelengths specific to that element: its characteristic "line spectrum." The technique could be highly sensitive—even tiny amounts of a metal salt might be enough to produce a detectable "fingerprint" spectrum.

Kirchhoff and Bunsen used their spectroscope to search for metals in all kinds of substances, looking for their unique spectral emission lines when the materials were placed in the flame. They studied minerals and mineral waters, spotting elements such as sodium, potassium, calcium, and iron.

As well as providing a means of analyzing substances to find out which known metals they contained, the spectroscope could reveal the presence of previously unknown ones, if it generated spectral lines that didn't match any of those catalogued for known elements. In 1860 the two scientists saw some blue emission lines in the spectra from the residue of evaporated natural mineral water that did not seem to match those of elements already recognized, and which they suspected of being from a new element in the same family as lithium, sodium, and potassium: an alkali metal. By the following year, they had managed to extract a tiny amount of a salt of this element from vast quantities of the mineral water, and confirmed

that it generated "two splendid blue lines situated close together." They felt justified in claiming the discovery of a new element.

A red line

"The bright blue light of its incandescent vapour," they wrote, "has induced us to propose for it the name *caesium*, derived from the Latin 'caesius,' used to designate the blue of the clear sky." (Today the standardized spelling is cesium.) It is indeed an alkali metal.

BELOW: Still from an animation of rubidium-cesium spectroscopy, 2016, by Rhys Lewis (animate4.com). Here, aqueous ions of cesium (Cs+) and rubidium (Rb+) are producing characteristic purple-blue and red colors in a flame test, with the resulting emission spectra also shown.

There was more. Studying a mineral from Saxony called lepidolite, the two scientists extracted another new salt which produced unknown spectral lines in the violet, red, yellow, and green parts of the spectrum. Of these, the line in the red region was one of the most prominent, prompting the pair to propose the name *rubidium* from the Latin for a deep red color. Because rubidium is rather rare and reactive, this alkali metal was not prepared in pure form until 1928.

OPPOSITE: Robert Bunsen and Gustav Kirchhoff's "Spectra of the Metals of the Alkalies & Alkaline Earths." From *Spectrum Analysis. Six Lectures Delivered in 1868 Before the Society of Apothecaries of London*, 1885, Science History Institute, Philadelphia.

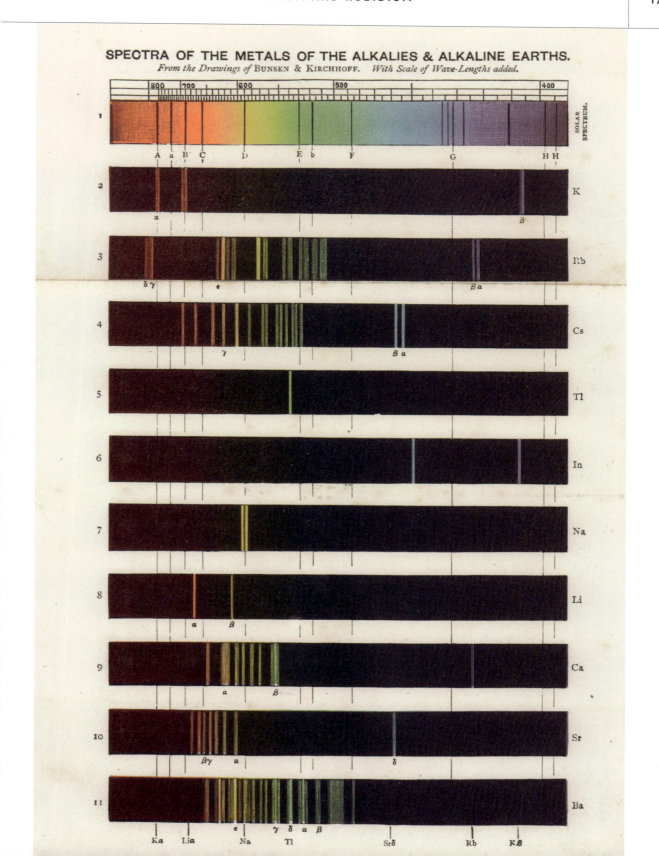

SPECTRA OF THE METALS OF THE ALKALIES & ALKALINE EARTHS.

From the Drawings of BUNSEN & KIRCHHOFF. *With Scale of Wave-Lengths added.*

THALLIUM AND INDIUM

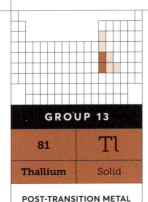

GROUP 13	
81	Tl
Thallium	Solid

POST-TRANSITION METAL

ATOMIC WEIGHT: 204.38

GROUP 13	
49	In
Indium	Solid

POST-TRANSITION METAL

ATOMIC WEIGHT: 114.82

William Crookes was one of the most colorful figures of British science in the Victorian era. His portfolio was wide and diverse: he trained at the Royal College of Chemistry, and set himself up as a chemical consultant, but he also wrote and published a magazine, *Chemistry News* (from 1859), pursued a close interest in the new technique of photography, and wrote on topics ranging from public sanitation to gold mining. He became very active in Spiritualist circles: ostensibly a skeptic who advocated the use of scientific techniques to distinguish genuine "spirit phenomena" from fraud, he was rather credulous of the claims of mediums and was duped into believing in their powers to communicate with and induce manifestations of the spirits of the dead.

Despite that rather disreputable passion, Crookes was highly regarded as a scientist. In 1898 he became president of the British Association for the Advancement of Science, having received a knighthood the previous year.

Much of that reputation rested on the discovery that Crookes made using a spectroscope like that invented by Gustav Kirchhoff and Robert Bunsen. Working in the laboratory he had fitted out in his London home, he found a new element. Crookes was sure that there were, in fact, plenty of new elements "waiting to be found out" using the spectroscope—and he was determined to find some. "I have seen *several* suspicious looking spectra," he told his collaborator Charles Williams in early 1861. He studied everything he could lay his hands on—including a rather unlikely source: the sludgy residue left over from the industrial production of sulfuric acid. The stuff was known to contain the element selenium, which is often present in natural sulfur. Crookes placed some of it in his spectroscope, and was surprised to see a green emission line not previously known: the signature of a new element.

Mindful of how Kirchhoff and Bunsen had named their two new metals after the hue of their spectral lines, he proposed

RIGHT: Sir William Crookes in his laboratory, 1890s, Wellcome Collection, London.

RIGHT: Thirty-one specimens illustrating the discovery of thallium by Sir William Crookes, ca. 1862, Science Museum, London.

the name *thallium*, derived from the Greek word *thallos* for a budding twig. "The green line which it communicates to the spectrum," Crookes wrote, "recalls with peculiar vividness the free colour of vegetation at the present time" (that being the spring).

The question was, though: was that enough to support the claim of a new element? Many chemists at the time considered that, before it could be added to the list of elements, you needed to have isolated the pure element itself in sufficient quantities to study its chemistry. That, after all, was why Bunsen and Kirchhoff had worked so hard to extract their cesium from mineral water. Crookes set Williams to that difficult task, pleading that he was too busy with "literary work" to be able to undertake it himself. In the end he could not wait, and despite Williams's advice that he claim only the detection of a new spectral line, Crookes asserted in his *Chemical News* in March that he had indeed found an element, probably belonging to the same group of the periodic table as sulfur and selenium. He didn't, however, have a sample of a thallium salt until January 1862.

He proudly exhibited the sample at the international exhibition at Hyde Park, in London, that May—only to be dismayed to learn in June that a French chemist named Claude-August Lamy had turned up with Antoine-Jerôme Balard, the discoverer of bromine, claiming to have an ingot of solid thallium.

The next year, two German scientists, Ferdinand Reich and Hieronymus Richter, went looking for this new element in zinc ore, but saw instead a new indigo-blue line. They declared that they had found another new element, which they named indium, and went on to obtain a pure sample of it. Although it was Reich who did the initial work, he was color-blind and so needed to enlist Richter's help to inspect the spectral line. So he was dismayed to find Richter claiming in 1867 to have made the discovery himself.

Indium sits above thallium in the periodic table, which is why they are chemically rather similar. Indium is one of the softest metals known: like sodium, it can be cut with a knife, and it melts at just 312.8°F, about the temperature at which melted sugar turns to toffee.

HELIUM

2

He

Helium

Noble gas

ATOMIC NUMBER
2

ATOMIC WEIGHT
4.003

PHASE AT STP
Gas

In 1802 the English chemist William Hyde Wollaston (who we encountered earlier in the discovery of palladium) repeated Newton's experiment on the splitting of sunlight into its spectrum. But he had better optical instruments than Newton, and noticed something new: there were gaps in the spectrum, dark lines where the light seemed to have been stripped out.

In 1814 the German scientist Joseph von Fraunhofer made the same observation independently, and used his better lenses to map out all the missing spectral lines, of which he counted more than 570. He designated these with letters: A through to K, with subscripts for apparent families of lines.

Where these gaps came from was not clear, however, until the work of Robert Bunsen and Gustav Kirchhoff. They realized that the "Fraunhofer lines" in the solar spectrum appeared at the same wavelengths as some of the emission lines that they saw in their spectroscope. They supposed that the corresponding elements in the solar atmosphere—or indeed in the Earth's atmosphere—were absorbing the light. In other words, here was a way to find out what the Sun consisted of.

A journey to India

It was also possible to see emission from these elements in the solar spectrum, as they re-radiated the light they absorbed. In particular, there were two strong lines in the yellow region, designated D_1 and D_2 by Fraunhofer, that corresponded to sodium. These were so intense, however, that it was hard to see any other fainter emission lines. That's why the French astronomer Pierre

BELOW: Solar spectrum shown by Fraunhofer lines, 1814, Deutsches Museum, Munich.

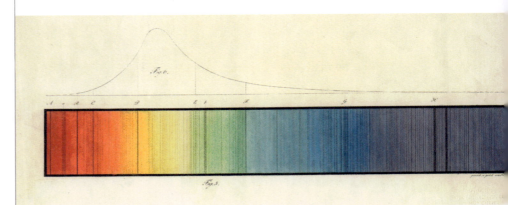

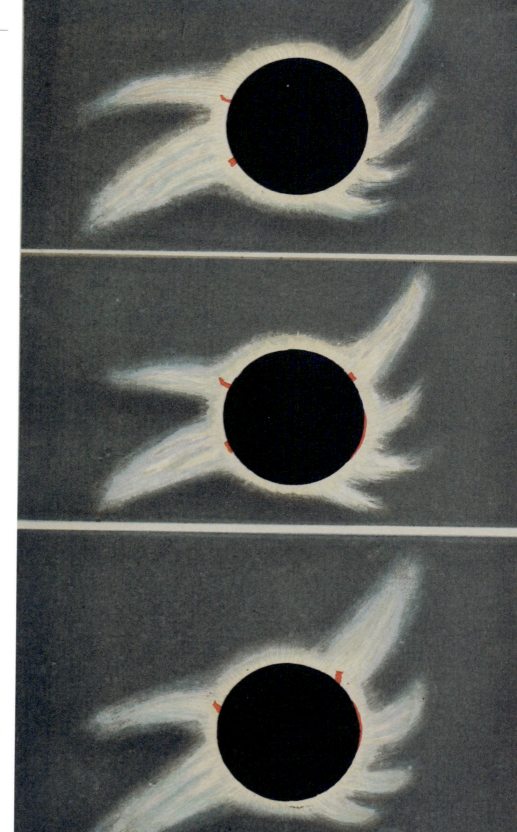

RIGHT: Sketch of the 1868 solar eclipse by M. Stephan. From *Archives des Missions Scientifiques et Littéraires*, Paris: Ministère de l'Instruction Publique, 1868, Vol. 5, Natural History Museum Library, London.

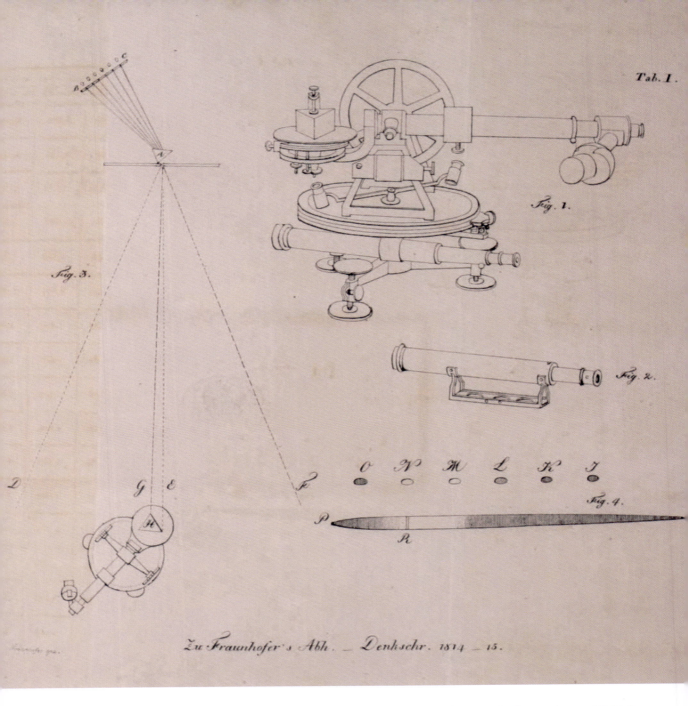

Tab. I.

Fig. 1.

Fig. 2.

Fig. 3.

Fig. 4.

Zu Fraunhofer's Abh. — Denkschr. 1814 — 15.

ABOVE: Joseph von Fraunhofer's spectroscopy. From his "In Relation to the Perfection of Achromatic Telescope," *Denkschriften der Königlichen Akademie der Wissenschaften zu München*, 1814, Vol. V, Natural History Museum Library, London.

Jules Janssen traveled to India in 1868 to take measurements of the solar spectrum during a total eclipse. He hoped that he might spot the emission lines of other elements in the spectrum of light from the corona during the eclipse. And, indeed, he saw another bright yellow line, which he thought might come also from sodium. But later that same year, the British astronomer Norman Lockyer measured the solar spectrum of the rather dim sunlight filtering through London's polluted and overcast skies, and he too saw the third yellow line, which he

designated D_3. After discussing the finding with the chemist Edward Frankland, he concluded that the new emission line must, in fact, come from a hitherto unknown element, present in the Sun. They named it helium, after the Greek sun god Helios.

Gas from a rock

This was quite an outlandish proposal: that the Sun contained an element not present on Earth. But was it really not? In 1882 the Italian physicist Luigi Palmieri was analyzing lava from an eruption of Mount Vesuvius using spectroscopy when he saw an emission line at the same wavelength as Lockyer's D_3 band. He figured that the lava must contain the solar element helium.

Yet in order to truly believe in this putative new element, chemists wanted to have a sample of it so that they could study its properties. The first isolation of helium seems to have been achieved by an American geologist named William Hillebrand, although he did not realize it at the time. In 1891 he dissolved a uranium mineral called uraninite in acid and saw that gas bubbles came from it. He used spectroscopy to examine the gas, but couldn't identify all its spectral lines. Because the gas was unreactive, he thought it might be nitrogen. In 1895, however, Per Teodor Cleve and Nils Abraham Langer at Uppsala University repeated the experiment and showed that the uraninite contained helium.

In the same year, Hillebrand's discovery came to the attention of the Scottish chemist William Ramsay, working at University College London. He got hold of some uraninite and asked his student, named Matthews, to repeat the experiment. Ramsay initially thought that the gas might be some new element, which he proposed to call "krypton," from the Greek for "hidden." But he found that the gas collected emitted a bright yellow line, suspiciously like that of helium. "I…began to smell a rat," he declared later. Because his spectroscopic equipment was rather poor, he sent samples of the gas to Lockyer and to William Crookes for a more accurate analysis. Crookes confirmed a day later that there

ABOVE: William Hillebrand, ca. 1900, Williams Haynes Portrait Collection, Science History Institute, Philadelphia.

was helium in it. As chemists gathered more of this gas, they were able to deduce that its atomic weight—a measure of the mass of its atoms—was very small: only hydrogen weighed less. Uranium, in contrast, was the heaviest element then known. So what was helium doing in a uranium mineral? The answer turned out to be more remarkable than anyone could have guessed: the helium was emitted by the very *nucleus* of uranium atoms, in the process that was soon to be called radioactive decay.

THE INERT GASES

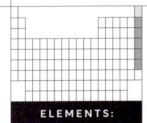

We saw earlier how, in the eighteenth century, burning objects were thought to release the element called phlogiston into the air; burning ceased when the air was totally "phlogisticated," or saturated with phlogiston. We now know that, on the contrary, burning stops when all the oxygen has been used up, and all that remains is nitrogen. Although nitrogen is rather unreactive, Henry Cavendish showed that this gas (which he considered to be phlogisticated air) can also be consumed in reactions with oxygen, induced by electrical sparks, that convert it to a kind of acid (actually to nitrogen oxides that will form nitric acid in water).

But Cavendish, who was a diligent observer and measurer, reported that he could never quite get rid of all the phlogisticated air this way. There was always a tiny bubble left that would not react at all, constituting about a 1/120th part of "common air." It was an unexplained puzzle, mentioned in a nineteenth-century biography of Cavendish by George Wilson.

In the late nineteenth century, William Ramsay bought a copy of Wilson's book while he was still a young chemistry student, and read about Cavendish's curious bubble. Later Ramsay studied the oxides of nitrogen that Cavendish had made, and became an expert in the chemistry of gases. Then, when he heard the distinguished scientist Lord Rayleigh (John William Strutt) speak in April 1894 on his studies of nitrogen, some memory of Cavendish's discrepant observation seemed to creep back into his mind. Rayleigh said that the density of nitrogen as measured by extracting it from air (that is, removing all the other known components) seemed different from that measured when nitrogen was produced chemically. Ramsay talked to Rayleigh afterward, wondering if perhaps there might be some hitherto unknown and very unreactive substance mixed in a tiny amount with nitrogen in air.

Ramsay repeated Cavendish's experiment and found that there was indeed an extra gas present, which Ramsay could not get to react, no matter what he tried. He supposed that it was a new, highly inert element, and proposed a suitable name for it: argon, from the Greek word for "lazy." We can't be sure, Ramsay wrote in 1896, that no elements will combine with argon, "but it appears at least improbable that any [such] compounds will be formed." That was not for want of trying; when Ramsay sent a sample to Henri Moissan, the French scientist attempted to get it to combine with the viciously reactive fluorine gas that he had isolated, but to no avail.

This peculiarly unresponsive element did not present much for the imagination to work on: when Ramsay exhibited a sample in a sealed glass tube at the Royal Society in 1895, the audience had to take it on trust that they weren't simply being shown a vial of plain old air.

Nonetheless, the discovery caught the attention of H. G. Wells, who began

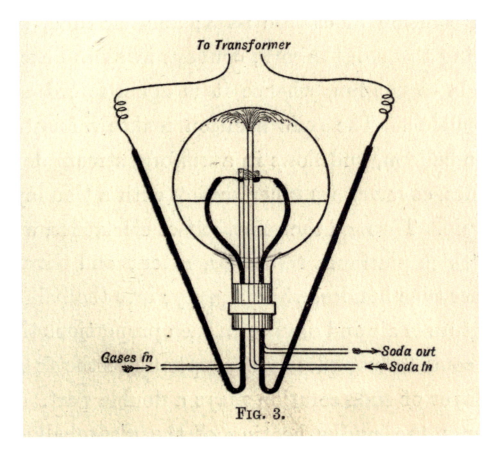

LEFT: Experiment with argon. From William Ramsay's *The Gases of the Atmosphere: The History of Their Discovery*, London; New York: Macmillan and Co., 1896, Fig. 3, University of California Libraries.

To Transformer

Gases in

Soda out
Soda in

FIG. 3.

to achieve literary fame after the publication of his "scientific romances" *The Time Machine* (1895) and *The Island of Doctor Moreau* (1896). In his sensationally successful 1898 novel *The War of the Worlds*, Wells described a poisonous gas used by invading Martians called "black smoke," chemical analysis of which using the spectroscope showed "the presence of an unknown element with a brilliant group of three lines." It is possible, the narrator added, that this element has the unique power to "combine with argon to form a compound which acts at once with deadly effect." Most of his readers had probably never heard of argon—but to a man of science like Wells, it was an exciting development in the story of the elements.

Argon, however, was not alone. Ramsay suspected that there might be other such elements still to be discovered—an entire new column of the periodic table, unknown previously because of their refusal to combine with other elements. We saw earlier that in 1895 he found helium in the mineral uraninite (also known as cleveite). But he had been hoping to find something entirely new there, and was frustrated.

Working with Morris Travers at University College London in early 1898, they turned the quest toward that tiny unreactive fraction of air. Using the techniques recently developed for liquefying air, they used "fractional distillation" to trap the argon: letting all of the air slowly evaporate down to the final, densest fraction. Then they used chemical methods to extract any remaining nitrogen, and investigated the residue—mostly argon—with a spectroscope in order to look for emission lines that might signify other elements.

Kingston line of hills began, there was a fitful cannonade far away in the south-west, due, I believe, to guns being fired haphazard before the black vapour could overwhelm the gunners.

So, setting about it as methodically as men might smoke out a wasps' nest, the Martians spread this strange stifling vapour over the Londonward country. The horns of the crescent slowly moved apart, until at last they formed a line from Hanwell to Coombe and Malden. All night through their destructive tubes advanced. Never once, after the Martian at St. George's Hill was brought down, did they give the artillery the ghost of a chance against them.

And through this two Martians slowly waded, and turned their hissing steam-jets this way and that.

ABOVE: Wellsian "black vapor" (said to be a compound of argon) in the first illustrations (by Warwick Goble) for H. G. Wells's *The War of the Worlds*, serialized in *Pearson's Magazine*, London: Pearson, 1897.

Completing the noble family

Quickly enough they found one, distinguished by bright yellow-green emission lines, which they called by the name Ramsay had previously invoked when he had hoped to find something new in the "emanation" from uraninite: krypton. They figured that there should also be a lighter gas between helium and argon, filling that space in the periodic table. In June they found it: a gas that produced a

"blaze of crimson light," as Travers put it: "a sight to dwell upon and never forget." They named it for its novelty: neon, from the Greek for "new." That red glow was soon to be seen ablaze in gas discharge tubes used for store and advertising signage all over the world.

A month later, Ramsay and Travers had bagged yet another inert gas: xenon, made by fractional distillation of krypton. The name means "stranger" or "outsider," further testimony to the oddness of this group of elements. Finally, in 1908, Ramsay made radon, the heaviest of the natural inert gases, which owes its name to its radioactivity. Indeed, radon was first seen as a product of the radioactive decay of the element thorium by the New Zealander physicist Ernest Rutherford and his chemist

BELOW: Mark Milbanke's portrait of Sir William Ramsay, 1913, University College, London.

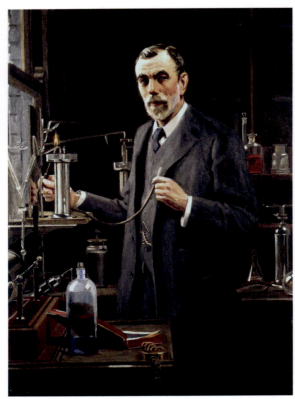

colleague Frederick Soddy, working at McGill University, in Montreal, in 1902. Here it appears in the process that converts one element to another by radioactive emission of particles from the atoms' nucleus. Radon from natural radioactive decay is present in some granites in sufficient quantities for its own radioactivity to be considered a health hazard as it leaks out of the rock.

Ramsay was awarded the 1904 Nobel Prize in Chemistry for his discoveries of the inert gases. And inert they have remained, for the most part: xenon and krypton can be induced to form stable chemical

ABOVE: The early impact of neon signage is evident in this photograph, "The Great White Way—Night Scene on Broadway above Times Square, New York City," Keystone View Company, 1928, Library of Congress Prints & Photographs Division, Washington, DC.

compounds, but argon will react only under rather exotic conditions—for example, when the atoms are ionized by intense radiation, or its very weak chemical bonds stabilized at very low temperatures. Neon and helium remain aloof from the other elements to this day.

RADIUM AND POLONIUM

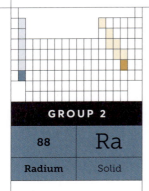

GROUP 2	
88	Ra
Radium	Solid

ATOMIC WEIGHT: (226)

GROUP 16	
84	Po
Polonium	Solid

ATOMIC WEIGHT: (209)

The "rays" that Henri Becquerel discovered coming from uranium in 1896 were a mystery. Able to trigger the darkening of photographic emulsion and yet invisible to the eye and insensible to the touch, they seemed rather like the X-rays that Wilhelm Roentgen had discovered a year earlier—and which had prompted the studies that led to Becquerel's finding. What was special about uranium that gave it this power to emit energy?

X-rays created a sensation in *fin de siècle* Europe because of their ability to reveal in a photograph dense objects (such as bones) that were hidden within or behind other materials (such as skin and flesh). But uranic rays (see page 100) were weaker, and didn't elicit the same fascination. One person who decided they warranted further attention, however, was a young Polish chemist looking for a subject for her doctoral thesis at the Collège de Sorbonne in Paris. "The question was entirely new," wrote Marie Curie later, "and nothing yet had been written upon it."

Curie, *née* Maria Skłodowska, came to study in Paris in 1891. There she met the French physicist Pierre Curie, who had already established something of a reputation in 1880 by discovering, with his brother Jacques, the phenomenon of piezoelectricity, in which some materials develop an electric field when squeezed. They married in 1895.

Marie Curie decided to investigate uranic rays in 1898, collaborating with her husband in a tiny laboratory at his workplace at the School of Chemistry and Physics. At first they investigated how uranium salts could, via their mysterious rays, induce an electrical charge in a nearby metal plate, which allowed them to measure the strength of the emission. But rather than rely on small

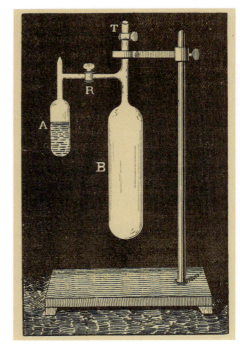

RIGHT: "Phosphorescence caused by the emanation of radium." From J. Danne's *Le Radium, Sa Préparation et Ses Propriétés*, Paris: Librairie Polytechnique Ch. Béranger, 1904, Fig. 33, Francis A. Countway Library of Medicine, Harvard.

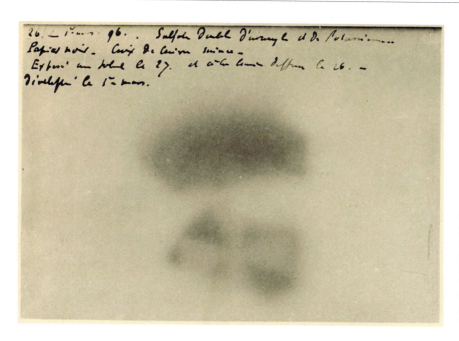

supplies of these substances (donated by Henri Moissan), Marie started to use raw uranium ore. And this was the odd thing: the ore could be an even stronger emitter than purified uranium. It was, in the new terminology they introduced, even more "radio-active."

The Curies reached a striking conclusion: the ore must contain some impurity that also showed radioactivity, but even more strongly than uranium itself. "I had a passionate desire to verify this new hypothesis as rapidly as possible," Marie wrote.

To do so meant separating and isolating this new radioactive source: a task that called for the kind of chemistry used to separate other new elements, such as the rare earth metals found in yttrium minerals. Typically this meant finding a reaction that precipitated one element as a solid while leaving others in solution. A new element could be precipitated, for example, if it was chemically similar to another that could be easily precipitated. A radioactive element had the advantage that you could always figure out "where it went"—into solution or into a precipitate—because the radioactivity went with it, detectable with the instruments the Curies had devised. As she worked, helped by the chemist Gustave Bémont, with solutions of uranium salts, Marie Curie found to

her surprise that there seemed to be *two* other radioactive sources in uranium ore: one that behaved rather like the element barium, the other like bismuth.

Little by little, the Curies used these extraction methods to prepare solutions that were far more radioactive than uranium. By July 1898, they

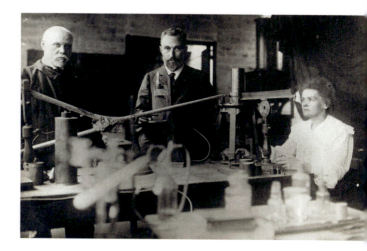

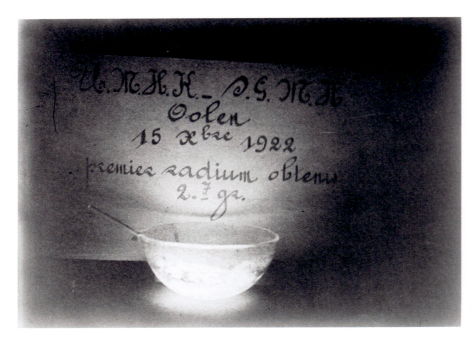

LEFT: A bowl containing the first radium bromide, photographed in the dark and lit by its own phosphorescence, 1922, Musée Curie; coll. Institut du Radium, Paris.

OPPOSITE: Marie Curie's notebook containing notes on radioactive substances, with sketches of apparatus, 1899–1902, Wellcome Collection, London. Curie's notebooks are said to be still radioactive themselves because of exposure to the substances she studied.

reported to the Institut de France that they had extracted from uranium ore "a metal never before known, akin to bismuth…If the existence of this metal is confirmed, we propose to call it *polonium* after the country of origin of one of us." Poland was ruled at that time by Tsarist Russia, and Marie was keen to assert its cultural independence.

Yet it was the other element, the one that "followed" barium, that they pursued first. By the time they had made a solution of it so concentrated that it was possible to identify in a new spectral line—the telltale signature of an unknown element —they found that the radioactivity was intense enough to make the water glow. That light inspired the name that Pierre recorded in his notebook close to Christmas of 1898: radium. "We had an especial joy," Marie wrote, "in observing that our products containing concentrated radium were all spontaneously luminous."

Marie labored to isolate ever purer samples of radium in a laboratory that was little more than an unheated shed on the premises of the School of Chemistry and Physics. But she didn't complain. "One of our joys," she later wrote, "was to go into our workroom at night, we then perceived on all sides the feebly luminous silhouettes of the bottles or

ABOVE: Marie and Pierre Curie in their laboratory, prior to receiving the Nobel Prize. From the cover of *Le Petit Parisien*, Paris: January 10, 1904, National Library of Medicine, Bethesda, Maryland.

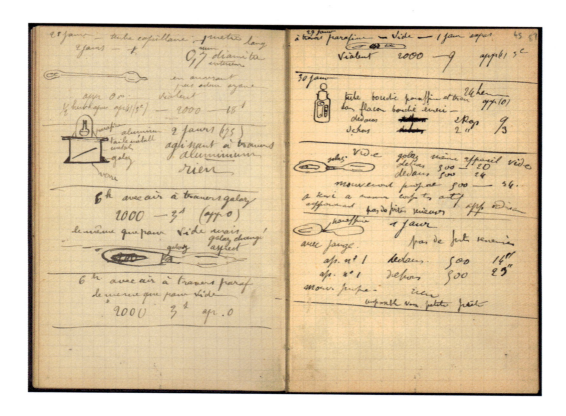

capsules containing our products. It was really a lovely sight and always new to us."

It took until 1902 before Marie Curie had what she needed to definitively claim the discovery of the new element: about one-tenth of a gram of a pure radium compound, from which she could measure properties such as the atomic weight. She submitted her doctoral thesis in June 1903, by which time the scientific world was buzzing with discussion of these new radioactive substances—and what radioactivity itself was. The emission seemed to persist inexhaustibly—where was all this energy coming from?

That same year, Marie and Pierre Curie were awarded the Nobel Prize in Physics for their contribution to understanding the phenomenon of radioactivity discovered by Becquerel (who shared the award with them). It was the first of two Nobels for Marie—the second, in chemistry in 1911, recognized the discovery and isolation of the two elements radium and polonium.

At first, the glowing radium was regarded as a miracle cure: salts of the element were sold as panaceas, radium was served in "glowing" cocktails, and radium paint was used for see-in-the-dark dials on watches and instrument panels. But by the mid-1910s, it was becoming clear that not only were the medical claims dubious, but radium might also be positively hazardous. Little by little it became obvious that radioactivity was a serious hazard to health—which explained why both Marie and Pierre often suffered from anemia, tiredness, and aching joints, and why their fingers might get inflamed and shed skin after handling flasks of the stuff. Toward the end of her life, Marie Curie researched the use of radium for cancer therapy, using its radioactivity to kill tumors. But it was too late for her herself: she died of leukemia, probably brought on by her work on radioactive materials, in July 1934.

CHAPTER EIGHT

THE NUCLEAR AGE

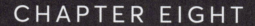

LEFT: The first H-bomb test, "Mike," conducted in Operation Ivy, on Elugelab Island, Enewetak Atoll, Marshall Islands, Pacific, November 1, 1952.

THE NUCLEAR AGE

T he advent of modern atomic science began as the nineteenth century drew to a close. No one was quite sure if it was physics or chemistry, but this new science both rationalized at last the nature and organization of the elements and shattered what had been comfortable certainties about the fundamental nature of matter.

It's ironic that atoms finally became accepted as real objects at the same time as their name—meaning "indivisible"—was shown to be a misnomer. In 1908 the French physicist Jean Perrin measured the motion of tiny grains of resin in water under the microscope, and showed that their erratic paths obeyed the mathematical rules predicted by Albert Einstein three years earlier. Einstein's theory was based on the idea that these "random walks" were caused by the impacts of water molecules, far too small to see, on the grains. So the result supported the view that matter is made up of such tiny particles, built from atoms. Perrin's 1913 book *Atoms* crowned this victory for the atomic theory, finally persuading most of the scientists who had until then remained sceptical, in the absence of any direct proof of the existence of atoms, that they were anything more than a convenient way of speaking.

However, it already seemed clear that atoms were not the smallest pieces of anything. In 1897 the British scientist Joseph John Thomson showed that the mysterious "rays" called cathode rays emitted from a negatively charged electrode in a vacuum tube were in fact made up of particles that bear a negative electrical charge. These were given the name "electrons," and they were recognized to be the constituents of electrical current. All electrons were identical regardless of the gas they came from, and so Thomson figured that they were components of the atoms of all chemical elements. They were the first *subatomic* particle.

What's more, it was soon discovered that the number of electrons an atom contains is the same as its atomic number, which defines its position in the periodic table. That number wasn't just a label for where the element came in the sequence from lightest to heaviest; it encoded something profound about the *structure* of the element's atoms.

More insights into the anatomy of the atom came from discoveries in radioactivity. Scientists deduced that some of the radiation that came shooting out of radioactive substances such as uranium were actually particles: little bits of the atoms. "Beta rays" proved to be "beta particles"—which, it seemed, were in fact nothing more or less than Thomson's electrons. And the New Zealander physicist Ernest Rutherford showed soon after the turn of the century that alpha rays were particles with a positive electrical charge. In an elegant experiment in 1908 at the University of Manchester, he showed that they were basically helium atoms stripped of their electrons.

Working with the chemist Frederick Soddy at McGill University in Montreal, Canada, Rutherford showed that when the radioactive element thorium emitted alpha particles, it seemed to be converted into a different element, which the two

researchers initially called thorium-X. That was unsettling, because the chemical elements were supposed to be immutable: the amount of them that nature provided was fixed, and there was nothing you could do to change it. But now it seemed one could *make* more of an element. To Rutherford, the idea that elements could be interconverted sounded dangerously like the discredited belief of the alchemists. All the same, that conclusion seemed hard to avoid.

ABOVE: Cathode-ray tube used by J. J. Thomson to discover the electron, 1897, Science Museum, London.

These scientists began to speculate about what atoms made from subatomic particles looked like. In 1902–1904, both Thomson and the Irish-Scottish physicist Lord Kelvin suggested that they were clouds of positive charge somehow studded with electrons, like plums in a plum pudding (then a staple dessert of the British dining table). But electrons, Thomson had shown, have only a fraction of the mass of a hydrogen atom. So, once it became clear a few years later that atoms had only their atomic-number's worth of electrons, the question became: where was the rest of the mass?

Rutherford answered that in 1909 when he and his students Hans Geiger and Ernest Marsden fired alpha particles (with all the heft of a helium atom, as he'd proved) like bullets at a thin gold foil. Most passed straight through, showing that atoms were mostly empty space. But some were deflected from their path, and a few bounced right back as though they had collided with some massive obstacle. Rutherford concluded that most of an atom's mass was concentrated into a very dense "kernel," to which he gave the Greek name *nucleus*. The nuclear age had begun.

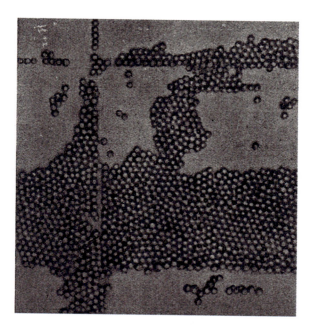

LEFT: Measurement of grains in a camera lucida. From Jean Perrin's *Atoms*, New York: D. Van Nostrand Company, 1916, University of California Libraries.

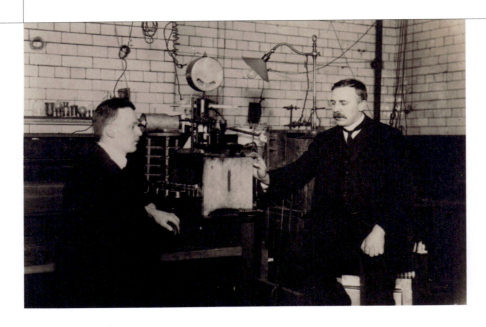

In 1911 Rutherford suggested that atoms consist of a nucleus with a positive electrical charge, surrounded by enough negatively charged electrons to balance it so that the atom is electrically neutral. It was rather like the way the planets orbit the (much more massive) Sun in the Solar System—held together in this case not by gravity, but by the forces of electrical attraction.

This "planetary atom" model was gradually refined in the following decades. The nucleus is itself a composite entity, made from other subatomic particles. One is the proton, which has an equal but opposite electrical charge to the electron, yet is almost two thousand times more massive. While electrons can be plucked from and added to an atom—this typically happens in chemical reactions—the number of protons remains constant, and is equal to the atomic number. *That* is the badge of identity of a chemical element: every atom of hydrogen, for example, has just one proton in its nucleus, and every atom of carbon has six.

Yet protons are not the only particles in the nucleus. All atoms (except the most common form of hydrogen) also contain particles with the same mass as a proton but no electrical charge: they are called neutrons, and were first discovered in 1932. Without neutrons in the mix, the positively charged protons in a nucleus would repel each other too strongly to hold together. The atoms of all elements can exist in different forms, called isotopes, which have different numbers of neutrons but the same number of protons. Hydrogen, for example, has three naturally occurring isotopes: the most abundant (99.98 percent of all hydrogen) has one proton and no neutrons, while hydrogen-2 (also called deuterium, and making up almost all the rest of the natural abundance) has one neutron, and hydrogen-3 (tritium) has two.

When radioactive elements decay by emitting an alpha or beta particle (and sometimes the third type of radiation, a gamma ray), the number of protons in their nucleus changes. These particles emerge from the nucleus; an alpha particle carries off two protons and two neutrons, thereby converting the atom to that of the element two spaces back in the periodic table. Even though beta particles are electrons, they come too, in this case, from the nucleus: a neutron splits into an electron, which gets spat out, and a proton, which remains behind. So the nucleus *gains* one proton and the element is converted to that one space forward in the table.

These processes of nuclear disintegration or *decay* are happening all the time in nature, at rates that vary not just from one radioactive element to another, but from one isotope to another. Some isotopes are stable more or less indefinitely, but others may decay at a rate defined by their so-called half-life: how long it takes for half of the atoms in a

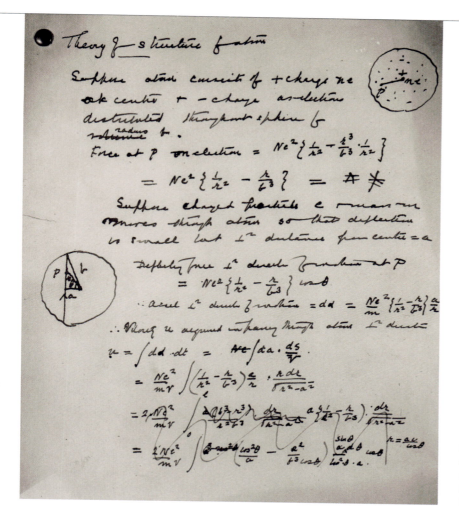

bunch of those atoms (a lump of the material, say) to decay. This half-life is always the same regardless of how many atoms you start with. The isotope carbon-14, for example, is constantly being produced in the Earth's atmosphere as subatomic particles streaming from space, called cosmic rays, collide with nitrogen atoms in the air. Carbon-14 has a half-life of 5,730 years. So when a plant or animal dies and the carbon-14 in its tissues stops being replenished by growth, the amount of the radioactive isotope it contains steadily decreases. By measuring that decline in its remains (a wooden carving or a fossilized bone, say), we can figure out how old it is. That's the basis of radiocarbon dating. Uranium-238, meanwhile, has a half-life of 4.47 billion years. It's one of many natural radioactive

elements in the Earth, and the energy of the radiation they emit by their gradual decay helps to keep the deep Earth hot and viscous, allowing the rock to keep churning and producing the slow process of continental drift.

At the same time that scientists in the early twentieth century began to understand the internal structure of atoms—and to figure out what it truly is that distinguishes one element from another—they also began to see how to induce and control these processes of nuclear reaction and decay. They acquired the power not just to liberate some of the energy bound up in the atomic nucleus, but also to induce nuclear reactions, converting one element to another—and to make new elements never before seen. The hunt for human-made elements was on.

TECHNETIUM

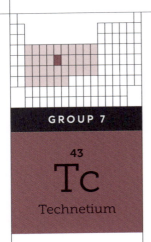

GROUP 7

43

Tc

Technetium

Transition metal

ATOMIC NUMBER
43

ATOMIC WEIGHT
(97)

PHASE AT STP
Solid

As the periodic table filled up with the discovery of new elements in the late nineteenth century, some obstinate gaps remained. One was below manganese and flanked by molybdenum and ruthenium, with an atomic number of 43. This missing element was evidently a transition metal, and there seemed to be nothing remarkable about it. It had even been predicted to exist by Dmitri Mendeleev himself when he drew up his periodic table. But no one could find it anywhere.

It wasn't for want of trying. In 1877 Russian chemist Serge Kern, working in St. Petersburg, claimed to have found a new metal that he thought belonged in this slot, which he called davyum after the English chemist Humphry Davy. But decades later it was shown to be a mixture of iridium and rhodium. And in 1908 the Japanese chemist Masataka Ogawa claimed to have found element 43, which he named nipponium after his country. He too was mistaken; he might have instead discovered rhenium (element 75), but even that isn't clear.

The discovery of rhenium is normally attributed to the Germans Walter Noddack and Ida Tacke (who later married) and Otto Berg, working in Berlin, who reported it in 1925. The Noddacks and Berg found this element in minerals such as ores of platinum and niobium (columbite). At the same time, they claimed to have seen evidence of another new element when they bombarded columbite with a beam of electrons. The material emitted weak X-rays that the scientists thought were a signature of a hitherto unknown element, and they figured it was the missing number 43. They proposed to call it masurium, after the region of Walter Noddack's homeland in East Prussia.

RIGHT: Ida Tacke and Walter Noddack in their laboratory at the Physikalisch-Technische Reichsanstalt (PTR), in Berlin-Charlottenburg, established in 1887, where rhenium was separated and characterized, Stadtarchiv Wesel, Germany.

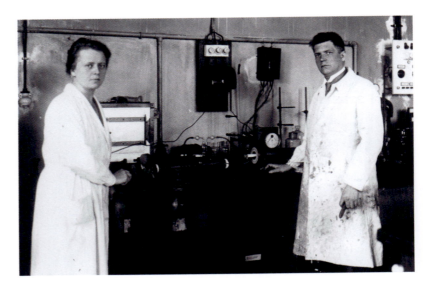

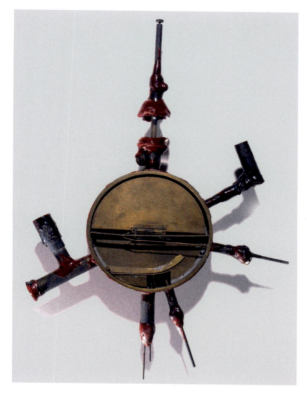

LEFT: The cyclotron, the first machine to accelerate particles to energies of more than one million electron-volts, Lawrence Berkeley National Laboratory, California.

Element 43 was not called masurium, however—because traditionally the discovery has been awarded to the great Italian radiochemist Emilio Segrè and his collaborator Carlo Perrier at the University of Palermo, in Sicily. It was made in 1937 by artificial transmutation, using a particle accelerator at the University of California at Berkeley to fire the nuclei of hydrogen-2 (deuterium) at a target of molybdenum metal.

This was cutting-edge nuclear physics at the time. Scientists had known since the early part of the century that they could induce nuclear reactions artificially—converting elements one to another—by colliding subatomic particles with atoms. At first the projectiles were particles emitted from radioactive atoms, such as alpha particles. In 1919 Ernest Rutherford bombarded nitrogen atoms with alpha particles from decaying radium, and he concluded that he could "disintegrate" the nitrogen nuclei this way: as the popular phrase had it, he "split the atom." (In fact, he hadn't really done that. Instead, his junior colleague Patrick Blackett showed that the nitrogen atoms hit by alpha particles had *gained* a proton, becoming oxygen.)

This approach, however, wouldn't work for heavier atoms, because their larger, more positively charged nuclei repelled the positively charged alpha particles before they could collide and get into the nucleus. To overcome that barrier, the bullets needed to have more energy than they acquired from radioactive decay. In 1929 the American Ernest Lawrence at Berkeley devised a machine for accelerating charged particles like this using electric fields, which he called a cyclotron (because the accelerated particles moved in cyclic spirals). Other researchers began to use the Berkeley cyclotron to see what nuclear transformations they could induce.

Segrè didn't himself do the transmutation experiments that made element 43. Having been a visitor at Berkeley, he was sent the plates of irradiated molybdenum to analyze using his chemical skills, to see if there was anything new

Identifying element 43

At the time their claim was discredited, but more recent experiments have shown that their samples of columbite could have contained tiny but detectable amounts of element 43 after all. It is created when uranium atoms decay by fission: they split apart spontaneously into much lighter nuclei, which can include this element. Columbite often contains appreciable amounts of uranium—up to as much as 10 percent. In 1999 David Curtis of the Los Alamos National Laboratory, in New Mexico, showed that the element does indeed appear in uranium ore, and he estimated that there might have been enough of it in the Noddacks' samples to have been detected. What's more, in the late 1980s the Dutch physicist Pieter Van Assche took another look at the results reported by the Noddacks and Berg, and argued that their case was more convincing than their contemporaries thought. Not everyone was persuaded, however, and, in all honesty, we can't say for sure who discovered element 43.

ABOVE: The 60-inch cyclotron in use at the University of California Lawrence Radiation Laboratory, Berkeley, in August 1939, Department of Energy, Office of Public Affairs, Washington, DC.

in them. That's how he and Carrier found the element that was named for being the first apparently made by technological means: it was called technetium. (Segrè's university at Palermo had wanted it to be "panormium," based on the Latin name for the town, but it was not to be.)

Why was technetium so elusive? Quite simply, it is radioactive—and even though the longest-lived isotope has a half-life of 4 million years, that's much too short for any significant quantities of technetium present when the Earth was formed four and a half billion years ago to still be in the ground today. This means that if we want any technetium today, we have to make it ourselves by nuclear transmutation.

There's one exception, however. In 1972 scientists discovered that a natural uranium deposit at Oklo in the Gabon, in Africa, had, around two billion years ago, become concentrated enough for the uranium to undergo slow spontaneous fission, turning the ore deposit into a natural nuclear reactor that burnt its "fuel" slowly for perhaps a million years or more. The nuclear process created small amounts of technetium—enough for some still to be detected in the Oklo minerals.

Despite its scarcity, technetium does have a use, and it's an important one. In 1938 Segrè, working with the nuclear chemist Glenn Seaborg at Berkeley, discovered that molybdenum-99 bombarded with neutrons decays into the isotope technetium-99 in a "high-energy" form said to be "metastable": this is denoted ^{99m}Tc. It sheds its extra energy in the form of gamma rays, becoming ordinary technetium-99, in a decay process with a half-life of six hours.

Each atom of ^{99m}Tc emits two gamma rays, which

can be detected and used to pinpoint their source. This makes ^{99m}Tc a kind of atomic beacon that is used to make medical images of the body. By attaching the atoms to molecules that will stick to particular tissues or cells, the gamma rays provide a map of those objects inside the body. For example, ^{99m}Tc-labeled proteins that stick to cancer cells are used to image tumors, and other molecules will attach the atoms to red blood cells to reveal the circulation, or to heart muscle to assess damage caused by heart attacks. ^{99m}Tc imaging has been used for a wide variety of organs and tissues: lungs, liver, kidneys, bones, and brain. Once ^{99m}Tc has decayed to ordinary ^{99}Tc, it is flushed out of the body in urine. The ^{99m}Tc used in these experiments comes from molybdenum-99 irradiated in a nuclear reactor, which is then shipped to hospitals where it decays to ^{99m}Tc over a period of several days. It's because of this valuable use that chemists have studied the chemistry of technetium—essential knowledge for attaching ^{99m}Tc labels to the right organs—much more intensively than you might expect for so rare an element.

BELOW: Emilio Segrè at the controls of the 37-inch cyclotron, June 12, 1941, Lawrence Berkeley National Laboratory.

NEPTUNIUM AND PLUTONIUM

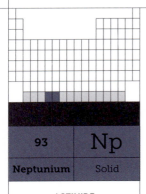

93	Np
Neptunium	Solid

ACTINIDE

ATOMIC WEIGHT: (237)

94	Pu
Plutonium	Solid

ACTINIDE

ATOMIC WEIGHT: (244)

Once scientists acquired the ability to induce and control nuclear reactions in the 1930s, they wondered if they could create entirely new elements not found in nature. The heaviest known element at that time was uranium, number 92 in the periodic table. Could the table be artificially extended beyond it?

The device used for these feats of nuclear alchemy was the particle accelerator, in which electric fields boost charged particles to very high energies. By smashing them into atomic nuclei, perhaps new configurations of protons and neutrons might be made? As well as Ernest Lawrence's cyclotron at Berkeley, the British scientists John Cockcroft and Ernest Walton at Cambridge built an accelerator for protons—that is, hydrogen nuclei. Unlike Lawrence's machine, theirs was "linear": it accelerated the particles in a straight line, not a spiral. Using this device in 1932, they fired protons at lithium atoms and saw them break apart into helium nuclei. Eight years later, Emilio Segrè, working with a team at the Berkeley cyclotron, fired alpha particles at bismuth (element 83) and discovered a hitherto unknown element, number 85, which they called astatine (the "unstable," because it decayed with a half-life of about seven hours).

Adding protons or alpha particles to nuclei wasn't the only way to make new elements. Neutrons, because they have no electrical charge, can get into a nucleus more easily than those positively charged particles. Discovered in 1932 by James Chadwick in England, they are emitted by radioactively decaying uranium atoms and

OPPOSITE: John Cockroft, a pioneer of particle accelerators, in the Cavendish Laboratory, Cambridge, 1932.

RIGHT: Edwin McMillan recreating the search for neptunium at the time of the announcement of the discovery, June 8, 1940, Lawrence Berkeley National Laboratory, California.

can be used as projectiles for transmuting other elements.

At first glance, absorbing a neutron doesn't seem to change the chemical nature of a nucleus. Its elemental identity depends only on the number of protons it contains; adding a neutron just makes it a different isotope. But when a nucleus undergoes beta decay, one of its neutrons splits into a proton and an electron (a beta particle), which gets ejected. The process also ejects an elusive, very light, and neutral particle called a neutrino. (There is also a type of beta decay that does the reverse, turning a proton into a neutron while emitting a positively charged "antimatter" electron called a positron.)

So beta decay—which happens in nuclei with a preponderance of neutrons—*increases* the atomic number of a nucleus by 1, turning it into the next element to the right in the periodic table.

Military secrets

In the 1930s there *was* no element to the right of uranium. Bombarding that heaviest of known elements with neutrons looked like a good bet for extending the periodic table into the virgin terrain. Segrè began to try this in collaboration with the nuclear physicist Enrico Fermi in Rome, in 1934. Later that year, Fermi and his colleague Oscar D'Agostino reported that they had seen evidence of two new elements made this way, with atomic numbers 93 and 94. They even proposed names: ausenium and hesperium. But the claim was soon shown to be wrong. In fact, what they had found were products of uranium *fission*, the nuclei splitting into much smaller pieces. So they had inadvertently discovered this important process four years before Otto Hahn and Fritz Strassmann reported it in Berlin.

By 1939, both Segrè and Fermi had been forced to flee fascist Italy for the United States. Segrè went to Berkeley, where he and Edwin McMillan fired neutrons at uranium to try to make "trans-uranic" elements. Element 93 should sit in the same periodic group as rhenium, and so they figured it would have similar chemical properties. But the element they identified seemed more akin to the rare-earth metals called the lanthanides, and they figured they had merely isolated one of those in their sample.

But they hadn't. In 1940 McMillan gained the assistance of an able young physicist called Philip Abelson, who showed that McMillan's uranium target did indeed contain element 93. Since uranium was named after Uranus, it seemed natural to name this new substance for the next planet out: it was christened neptunium. It isn't a wholly artificial element, existing in very tiny amounts in natural uranium ores, where some uranium atoms are transmuted by the neutrons emitted by others. But McMillan and Abelson were the first to see it.

Could we go further? In 1941 the Berkeley team, led by the young chemist Glenn Seaborg, used the cyclotron to fire nuclei of hydrogen-2—deuterons, with one proton and one neutron—at uranium. The collisions generated a neutron-rich isotope of neptunium, which then decayed by beta emission to make element 94. Continuing the astronomical nomenclature, it was named plutonium, after Pluto—the outermost "planet," and also the god of the underworld.

Unlike neptunium, the discovery of plutonium wasn't reported straight away. By this time, the military potential of nuclear reactions as a source of tremendous energy was clear, and research like this was sensitive and classified information, no longer to be described in scientific journals. So the paper announcing the discovery of plutonium wasn't published until 1946. Seaborg and McMillan were awarded the 1951 Nobel Prize in Chemistry for their work on trans-uranic elements—in the same year Cockcroft and Walton were given the Physics Nobel.

The military secrecy was fully justified, since both the Allies and the Germans quickly realized after Hahn and Strassman reported uranium fission in 1938 that it could be used to make a bomb of unprecedented power. The Berkeley team, including Segrè, soon found that one plutonium isotope, plutonium-239, had a similar fissile nature to uranium-235, so that it too could be used in a bomb. But it had to be made artificially, by neutron bombardment of uranium. A plant for plutonium-

239 production was quickly built at Oak Ridge, in Tennessee, and by 1945 enough had been created (several kilograms) to make a trial nuclear bomb. This was detonated in the Trinity test in the New Mexico desert on July 16. The second plutonium bomb, code-named "Fat Man," was dropped on Nagasaki on August 9, killing around 70,000 people

ABOVE: Plutonium bomb Trinity Test, White Sands Proving Ground, New Mexico, July 16, 1945.

in the blast. The trans-uranic, human-made elements had arrived, and the advent was unimaginably terrible.

THE ACCELERATOR ELEMENTS

AMERICIUM, CURIUM, BERKELIUM, AND CALIFORNIUM

95	Am
Americium	Solid

ACTINIDE

ATOMIC WEIGHT: (243)

96	Cm
Curium	Solid

ACTINIDE

ATOMIC WEIGHT: (247)

97	Bk
Berkelium	Solid

ACTINIDE

ATOMIC WEIGHT: (247)

98	Cf
Californium	Solid

ACTINIDE

ATOMIC WEIGHT: (251)

Once it was possible to reach beyond uranium by sticking more protons and neutrons onto its nuclei, scientists realized that the process could be bootstrapped further. You make the early trans-uranic elements neptunium and plutonium, and then add more particles to *them*.

That's how Glenn Seaborg's team at Berkeley made elements 95 and 96 in 1944. The heavier of the two came first, created in the summer by firing alpha particles at plutonium-239; element 95 came after, made by adding two neutrons to plutonium. Both elements remained classified until the war ended, when they were given names. Element 95 began the practice of naming these artificial elements after great scientists—and who better to award that first honor to than the husband-and-wife team who pioneered the whole field of radiochemistry, Pierre and Marie Curie? It is curium. Element 96, meanwhile, was saddled with the nationalism that was the legacy of late-nineteenth-century naming practice, given more baggage now by the start of the Cold War: it was called americium.

Despite all the secrecy that had surrounded their creation during wartime, these two elements had what is probably the most casual public announcement ever. Seaborg mentioned them in November 1945 on a US children's radio show, in response to a listener's question, just a few days before the findings were presented in more sober fashion to the American Chemical Society.

The new radiochemists began to wonder about the new elements as chemical entities. How did they combine with other elements? And did their behavior sustain the trends and regularities for which the periodic table was named? Did they follow the same chemical logic as the natural elements? With the discovery of curium and americium, Seaborg began to discern an unexpected pattern to their chemical properties. They didn't

RIGHT: Glenn Seaborg (left) and Edwin McMillan, who shared a 1951 Nobel Prize for their work in radiochemistry: the study of radioactive elements, Lawrence Berkeley National Laboratory, California.

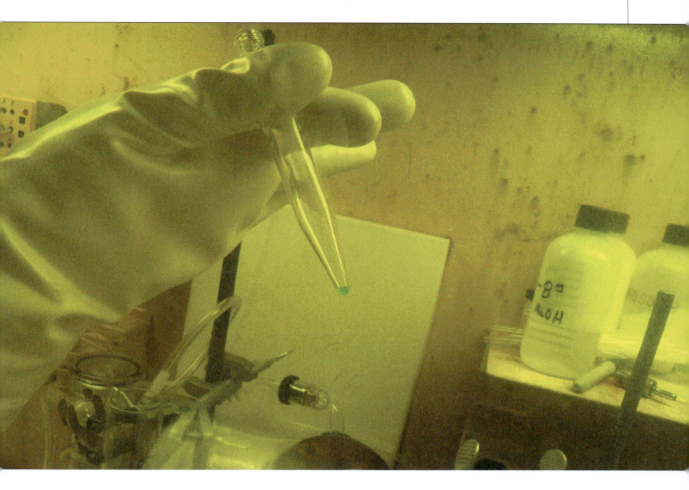

resemble the elements that seemed to sit above them in the periodic table: iridium and platinum, respectively. Instead, they turned out to form chemical compounds more like those of the lanthanide elements—the series of 14 elements squeezed in between lanthanum (element 57) and hafnium (element 72). Seaborg proposed that they are part of an analogous series that begins after actinium, with thorium (element 90). By analogy, then, he called them the actinide elements.

The Berkeley team, led by Seaborg and Albert Ghiorso, advanced to the next level of bootstrapping by bombarding americium and curium with alpha particles. In 1949 they made element 97 in this way; the next year, element 98 followed. They called the first of them berkelium, the second californium. The *New Yorker* magazine waggishly wondered why, since these were produced at the University of California at Berkeley, they had not teed up their

ABOVE: Vial with 13 milligrams of berkelium-249, made in the research reactor at the Oak Ridge National Laboratory, Tennessee, which was used to create the "artificial" element tennessine in 2009.

run more precisely by naming them "universitium" and "offium," so as to reserve the two new names for what followed. The Berkeley team responded that they'd have looked pretty foolish if some New York scientists had beaten them to it and instead christened the next two elements "newium" and "yorkium."

It was a joke with a pertinent point. The tradition of naming elements not necessarily for their country but for their region or town of discovery was, of course, an old one. But now the Berkeley team had instigated the idea that it was also *institutes* that could be recognized in element names: a triumphant "We were first" declaration. For, even if they did not

truly face serious competition in New York, they weren't the only runners in the race. Element-making was becoming an international sport, and, like most such competitions at that time, it was afflicted by Cold War tension and rivalry.

A narrowing window

Some of these new elements were fairly stable, so that they could be steadily accumulated and even isolated in quantities big enough to see with the naked eye. The isotope americium-241, one of the first to be made, has a half-life of 432 years, and for americium-243 it is 7,370 years. So this element may be created abundantly enough to find applications, most prominently as a source of gamma rays in smoke detectors. The gamma rays knock electrons from molecules in the air, ionizing them and letting a very small electric current pass between two electrodes in a circuit. If smoke particles enter the chamber and block the current, the alarm is triggered. Curium-242, the first of its isotopes to be made, has a half-life of just 160 days or so, but some heavier isotopes have half-lives running into the thousands or even millions of years.

By the time we reach berkelium, however, that stability is starting to wane. The first isotope to be made, berkelium-243, has a half-life of just four and a half hours, although berkelium-247 decays with a half-life of 1,380 years. And the first atoms of californium had a half-life of 44 minutes. Already it was becoming clear that, if chemists wanted to study these trans-uranic elements, the window of opportunity was going to narrow as the elements got more ponderous.

RIGHT: Edwin McMillan (left) and Edward Lofgren on the shielding of the Bevatron cyclotron, 1950s, Lawrence Berkeley National Laboratory, California.

THE BOMB-TEST ELEMENTS
EINSTEINIUM AND FERMIUM

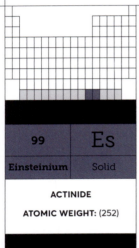

99	Es
Einsteinium	Solid

ACTINIDE
ATOMIC WEIGHT: (252)

100	Fm
Fermium	Solid

ACTINIDE
ATOMIC WEIGHT: (257)

The nuclear fission of uranium, discovered in 1938, showed that the immense amount of energy known for decades to be locked into the atomic nucleus could be harnessed and released on demand, if scientists could learn to control the fission process. By the end of the war it was understood how to build both a nuclear reactor, to release the nuclear energy slowly, and a nuclear bomb, which liberated it in one awful, obliterating burst.

Even more was possible—for better or worse. In 1919 the British physicist Francis Aston invented an instrument for measuring atomic weights of elements very accurately, and he found that the masses of the elements were not exactly integer multiples of the mass of the hydrogen atom, even though the hydrogen nucleus (a proton) was their building block; they were slightly lighter than that. (The other nuclear particle, the neutron, was not known at that stage.) Aston figured that the missing mass had been converted to energy, via Einstein's iconic relationship $E = mc^2$, when the nuclear particles came together to form heavier nuclei in a process called nuclear *fusion*. The mass decrease was very small, but all the same it meant that the energy liberated in fusion was enormous. "To change the hydrogen in a glass of water into helium," Aston wrote, "would release enough energy to drive the *Queen Mary* across the Atlantic and back at full speed."

Researchers soon figured that this process of nuclear fusion produced the energy of stars like the Sun: they are huge, dense balls of hydrogen that is being fused into helium. Every second, around 600 million tons of hydrogen in the Sun are converted to helium. That, however, isn't the end of the fusion process. Once a star has burned up most of its hydrogen into helium, it contracts and ignites the fusion of helium into still heavier elements, such as carbon and oxygen. At a later stage, those elements are also fused to make sodium, magnesium, silicon, and others. Stars, in other words, are the natural element factories, where all the natural elements are created by nuclear fusion.

Fallout

The nuclear scientists realized that there is potentially more energy to be harvested from nuclear fusion than from fission. To release it, hydrogen needs to be made tremendously dense and hot, which is not easy to do in a controlled fashion. Achieving the conditions that fuse hydrogen atoms in the Sun is quite impractical, but the isotopes hydrogen-2 (deuterium) and hydrogen-3 (tritium) will fuse under less extreme conditions. This is the process that, in 1942, Enrico Fermi and the Hungarian-American physicist Edward Teller realized could be used to make a "superbomb" many times more powerful than a uranium fission bomb. Teller implored the American government to pursue

the idea before the Nazis—and also the Soviets—figured out how to do it.

During wartime, the efforts of the Manhattan Project were focused on fission nuclear bombs, and so this "hydrogen bomb"—also called a thermonuclear bomb, because heat triggers the nuclear reaction—was not developed until afterward. The first H-bomb test, code-named "Mike," took place in 1952, on the Enewetak Atoll in the Marshall Islands of the Pacific. A thousand times more powerful than the Hiroshima bomb, it vaporized the islet on which it was detonated. Three years later, the Soviet Union carried out their first hydrogen-bomb test, marking the start of the nuclear standoff and the era of Mutually Assured Destruction.

Yet there was a novel scientific bounty from the Mike test. Filters from aircraft flown through the mushroom cloud, and coral from a nearby atoll, were sent to Berkeley for analysis of the debris from the nuclear fallout. (One of the jets that collected the filter samples lost its course when the electromagnetic pulse from the bomb scrambled the electronics and, running out of fuel, crash-landed in the ocean, killing the pilot Jimmy Robinson.) The radiochemists found evidence of two new elements, with atomic numbers 99 and 100. They named the first after the scientist whose famous equation had pointed the way to the H-bomb: einsteinium. The second honored Fermi's pioneering contributions in understanding and harnessing nuclear energy: it was called fermium. For security reasons the discoveries were not announced until 1955, and, although both Einstein and Fermi were still alive when the names were proposed, neither lived to see the official announcement of the new elements.

Towards "superheavy" elements

What were such heavy elements doing in the fallout of a hydrogen bomb? They were produced from the uranium used in the fission bomb that was detonated as the fuse to ignite the fusion of deuterium and tritium, as the uranium atoms were doused with neutrons. In 1954 the Berkeley team reported that they had made the two elements in

ABOVE: Enrico Fermi at the blackboard, ca. 1940s, US Department of Energy, Washington, DC.

the laboratory by irradiating plutonium and californium with neutrons.

Several months earlier, a team at the Nobel Institute for Physics in Stockholm, Sweden, also made fermium. Their approach marked a new way of creating these trans-uranic heavy elements. Instead of trying to boost the atomic number one or two at a time by planting neutrons or alpha particles onto the nuclei of a target, they added a substantial new chunk to uranium nuclei by firing ions of oxygen at them in a particle accelerator. The Berkeley group too were exploring that method, which could jump many steps forward along the row of trans-uranic elements in a single bound. This fusion of two relatively heavy nuclei was to become the key means of making new "superheavy" elements in the decades to follow.

THE EARLY TRANS-FERMIUM SUPERHEAVY ELEMENTS

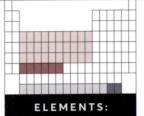

ELEMENTS:

Mendelevium 101

Nobelium 102

Lawrencium 103

Rutherfordium 104

Dubnium 105

Seaborgium 106

Bohrium 107

Hassium 108

When Swedish scientists reported that they had made element 100 in 1954 by bombarding uranium with oxygen ions, they seemed like parvenus to the experienced element-hunters in the big American and Soviet labs. Yet despite their relatively cheap and inferior resources, they were determined contenders. In 1957 the Stockholm team presented evidence for having made element 102, for which they proposed the name nobelium, after Alfred Nobel, the chemist who established the prizes. Their data were shaky, though; the claim wasn't believed at the time by others, and has never been confirmed.

In any event, the Soviet group had already staked a prior claim. The Soviet center for element-making was the Joint Institute for Nuclear Research (JINR) in Dubna, near Moscow, one of the Soviet state's science hubs, which was led by nuclear physicist Georgy Nikolayevich Flerov, a veteran of the Soviet nuclear-bomb program. In 1956 Flerov's team said they had made element 102 by firing oxygen ions at plutonium; they proposed to call it joliotium, after Marie Curie's daughter Irène Joliot-Curie, who had followed in her mother's footsteps and become a leading figure in nuclear science in the 1930s and 1940s, and her husband (and avowed communist) Frédéric Joliot.

In 1958 the Berkeley team asserted that *they* had the first convincing evidence of element 102, made by colliding carbon ions into a target containing the artificial element curium. This became the typical pattern for the next two decades or so: rival teams publishing evidence for a new element while casting doubt on that of their competitors. How could these claims be adjudicated? They were assessed then, and still are today, by the International Union of Pure and Applied Chemistry (IUPAC), which asks a panel of experts to weigh the evidence. During the Cold War, however, even this international body became mired in contention. IUPAC approved the name "nobelium" for element 102 in 1957—and that has stuck, even though the Dubna claim of 1956 was eventually awarded precedence over the Swedish one. But by the 1980s, these disputes had created chaos and confusion in the farthest reaches of the periodic table.

Take element 104. The Soviets said they had made it in 1964 by fusing plutonium and neon ions, and they called it kurchatovium after Igor Kurchatov, head of the Soviet nuclear science program, who masterminded the first Soviet nuclear bomb. But Al Ghiorso at Berkeley and his team contested that claim and insisted that they had found the first compelling evidence for element 104 from collisions of californium and carbon in 1969. They called it rutherfordium, after Ernest Rutherford, the discoverer of the atomic nucleus. The Soviets used their name, and the Americans theirs, and soon the literature of these "transfermium" superheavy elements was in disarray.

It only got worse. Flerov's team at JINR reported evidence for 105 in 1967, and three years later they suggested the cumbersome name nielsbohrium,

ABOVE: Updating the periodic table: Albert Ghiorso adds "Lw" (the initial abbreviation for lawrencium) to space 103, alongside codiscoverers Robert Latimer, Dr. Torbjorn Sikkeland, and Almon Larsh, 1961. Photograph by Donald Cooksey, The US National Archives.

after the Danish physicist Niels Bohr who, in 1912, showed how the theory of quantum mechanics might explain the way electrons were arranged in atoms. Predictably, Ghiorso and the Berkeley group presented their own case three years later, proposing the name hahnium after Otto Hahn. It was the same story for element 106—and for 107 there was yet another petitioner. In Darmstadt, Germany, the Laboratory for Heavy Ion Research (known by the German abbreviation GSI) had entered the race with a purpose-built particle

accelerator for colliding heavy ions (such as calcium and chromium) with targets such as bismuth. This represented another new strategy: not adding small lumps like carbon nuclei to heavy ones like uranium, but merging two nuclei of intermediate

size. The GSI team said they had made element 107 in 1981, but the Dubna group claimed to have produced it five years earlier.

In 1985 IUPAC, in collaboration with the International Union of Pure and Applied Physics (IUPAP), set up a Transfermium Working Group to assess the various claims for elements 104–107. The committee announced its decisions in 1992, declaring that in some cases a clear-cut decision about priority wasn't possible: as in all of science, sometimes a result has to be considered plausible, but not definitive. Still, they had to assign names. In 1994 the Working Group ruled that element 104 was to be called dubnium in recognition of the Soviet group's efforts, 105 was joliotium, and 106 acquired the name previously proposed for 104: rutherfordium. Element 107 was bohrium, and 108 hahnium.

Immediately there was dissent. The German team who were awarded priority for element 108 didn't want to call it after Otto Hahn, but insisted instead on their preference hassium, after the German state of Hesse in which GSI was situated. Even more controversially, the Americans at Berkeley—where the lab was now called the Lawrence Berkeley National Laboratory—had started in 1994 to refer to element 106 as seaborgium, after Glenn Seaborg. Fair enough, you might think: no one disputed the immense contributions to the field that Seaborg had made. The problem—if one can put it that way—was that he was still alive.

Continuing controversy

No element had previously been named after a living scientist. True, there was not exactly a "rule" about it—both einsteinium and fermium had been proposed while the two scientists were still alive, after all. But IUPAC seemed to have decided that this was now the tradition—until, faced with a rebellion by the American Chemical Society, they relented. And so, in 1997, the names were redistributed again: 104 was rutherfordium, 105 dubnium, 106 seaborgium.

The whole business has been dubbed the Transfermium Wars, and it does nuclear chemistry little credit, seeming to reveal the field as a hotbed of nationalism, chauvinism, triumphalism, and egotism. Besides, all the arguments about priority and naming risked eclipsing the really important questions, which were about what these elements were like *chemically*. The typical lifetimes of these "superheavy" elements got ever shorter with their increasing mass, and their rate of formation got ever slower. So it required ever more skill and ingenuity to find answers.

Take seaborgium. The typical production rates at GSI were just a few atoms per day, while the half-lives of even the longest-lived isotopes known in the 1990s are measured in seconds. (The record now stands at 14 minutes for seaborgium-269, reported in 2018.) Nonetheless, the GSI scientists devised an experimental system for very rapidly separating seaborgium atoms from the rest of the debris of the collisions used to make them, and carrying them along tubes in a flow of gas into a chamber where the handful of atoms could be reacted with chemicals like oxygen to form a compound that could be rapidly analyzed—all within a few seconds. It's the very fact that the seaborgium atoms decay that allows them to be seen at all in these experiments, from the alpha particles they emit at a characteristic energy. In this way, researchers have been able to figure out the composition of seaborgium's compounds, as well as properties like solubility. They have also succeeded in studying bohrium and hassium this way—but beyond that, the half-lives are generally too short to allow enough time to deduce much at all.

What is driving this determination to find out about the chemistry of the superheavy elements is the question of whether the periodic system of elements still works at all in this exotic regime of artificial, superheavy elements. Might this much-vaunted scheme for organizing the elements start to break down at such extremes?

OPPOSITE: Glenn Seaborg at a blackboard noting trans-uranic elements at the Lawrence Berkeley National Laboratory, California, November 1951, The US National Archives.

THE END OF THE ROW

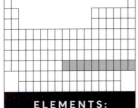

ELEMENTS:

Meitnerium 109

Darmstadtium 110

Roentgenium 111

Copernicium 112

Nihonium 113

Flerovium 114

Moscovium 115

Livermorium 116

Tennessine 117

Oganesson 118

OPPOSITE: Structure of the linear accelerator UNILAC (UNIversal Linear ACcelerator), used to make new elements at the GSI Helmholtz Center for Heavy Ion Research, near Darmstadt, Germany.

When the Cold War thawed, so did the Transfermium Wars. The search for new superheavy elements today happens in a much more collaborative spirit than it did up until the 1990s. This isn't just a reflection of the shift in geopolitical relations, though—it's more or less a necessity brought about by the sheer difficulty of the task. Extending the periodic table beyond element 108 or so is extraordinarily hard, and teams from different countries need to help one another: to share samples, check and verify claims, and offer expertise.

What's more, the game is about much more than getting to the next superheavy first. Faced with such a fearsome challenge, the teams are focusing increasingly on consolidating what we already know: making known superheavy elements in greater quantities so that their properties can be studied, and also trying to develop a better understanding of what it is that determines the stability of these swollen, fissiparous atoms.

Between 1981 and 1996 the team at the Center for Heavy Ion Research (GSI) in Germany made all the elements between 107 and 112. The last of these was named copernicium, after the astronomer Nicolaus Copernicus who proposed the Sun-centered model of the cosmos in the sixteenth century. (It was a curious choice, since Copernicus had nothing to do with atoms or chemistry.) The element was first seen in 1996 from collisions of a beam of zinc ions with a lead target; this fusion of two medium-sized nuclei of similar mass is called the "cold fusion" method, because it requires less energy to make the nuclei merge than if you start with an already very heavy element and try to add a much smaller nucleus onto it. The discovery wasn't officially confirmed until 2009, after plenty of further studies confirming the claim.

Element-makers internationally have now succeeded in the tremendous feat of finishing off the entire bottom row of the periodic table, which brings the total count to 118. The last of these sits at the foot of the column of inert gases that begins with helium. It is called oganesson, after Yuri Oganessian, the scientific leader of the Russian group at JINR in Dubna who first identified it in 2002—making it only the second element after seaborgium to be named for a living scientist. The first detection, from the fusion of californium with calcium ions, involved just one or two atoms of the new element, which underwent alpha decay with a half-life of just 0.69 milliseconds. The discovery was confirmed in 2006 in a collaboration between the JINR and American scientists from the Lawrence Livermore National Laboratory in California.

Such faint and fleeting sightings are hard to be sure of, or to confirm. In December 2015, the adjudicating committee of IUPAC and IUPAP declared that the Dubna/Livermore effort had reported convincing evidence for elements 115, 117, and 118. Element 115, first seen in 2003 at JINR,

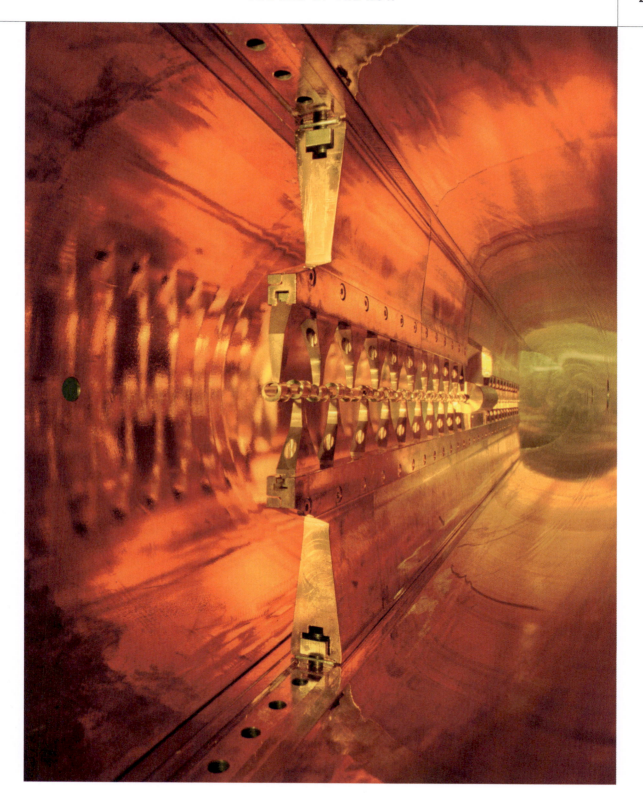

was named moscovium (Dubna being in the Moscow region). Element 117—the most recent such discovery to date—is called tennessine, for although it was also made at JINR, that experiment used a target of berkelium produced at the Oak Ridge National Laboratory in Tennessee. The berkelium sample—22 milligrams in all, completed in December 2008—had a half-life of just 330 days, ninety days of which were used up in purifying the artificial element. Then it had to be shipped as quickly as possible to Dubna for the collider experiments. As you might imagine, sending a highly radioactive international package (sealed in a lead container) requires some rather stringent paperwork—and on the first flight from New York to Moscow the documentation was left behind, so it was shipped back. On the second attempt, Russian customs officials found a problem with it, and back it went again over the Atlantic. It made the trip five times in all—each time with fewer of the precious berkelium atoms—until it was allowed through. Even then, an over-zealous Russian customs officer wanted to open the package to check it, until he was persuaded that this would be a very bad idea. After 150 days of firing calcium ions at the material, the Russian-American team announced the sighting of six atoms of element 117—soon to be tennessine—in April 2010.

The IUPAC/IUPAP committee also declared that element 113 had been first made by a team at the RIKEN Nishina Center for Accelerator-based Science in Wako, Japan, in 2004. The Japanese group used the "cold fusion" approach pioneered at GSI (the theory was developed by Yuri Oganessian), here fusing zinc ions with a bismuth target. It was the first superheavy element to be discovered in Japan, and the RIKEN team named it nihonium after the Japanese name of their country.

One of the key questions about these extreme superheavy elements is whether they sustain the periodicity in chemical behavior that underpins Mendeleev's table. The trends in the table can be disrupted for heavy elements by the effects of special relativity—the theory devised by Einstein in 1905 that describes objects moving extremely fast. The energies, and thus the speeds, of the innermost electrons in these atoms may be so high, because of the strong electrostatic interactions with the highly charged nuclei, that their mass gets larger, just as special relativity predicts. This means the electrons are pulled farther in toward the nucleus, so that they more efficiently screen the outer electrons from the nuclear charge. This "relativistic effect" alters the energies of the outer electrons, and thus influences the atom's chemical reactivity. Relativistic effects are already evident in the properties of familiar elements—in the yellowish color of gold and the low melting point of mercury, for example. They have been discerned in the chemical behavior of transfermium elements like dubnium (element 105).

But finding out about the chemistry of the biggest superheavies is barely feasible, because they are made only an atom or two at a time and decay so quickly. All the same, a few clues have been gleaned. One relatively simple and quick technique measures how strongly atoms are absorbed from a gas onto solid surfaces. Experiments at GSI have shown, for example, that flerovium (element 114) is metallic like the element above it (lead), but less reactive, whereas nihonium (113) forms strong chemical bonds to a gold surface.

In these extreme atoms, the norms of chemical behavior may eventually break down entirely. How an element reacts depends on the way its electrons are arranged into shells. But when the nuclei are so massive, the shells themselves may start to blur, creating what looks to be an almost undifferentiated cloud of electrons that defies any simple description. That's what is predicted for oganesson. No experiments can yet tell us anything about its chemistry—the longest-lived isotope has a half-life of less than a millisecond—so scientists have to rely on predictions calculated from the equations of quantum mechanics. These imply that oganesson has a sloshy, smeared-out shroud of electrons, making it unlike the inert gases that lie above it in the periodic table. It should form chemical bonds more easily, and many oganesson atoms—if they could be made—should clump into a solid rather than remain aloof from one another as a gas.

Seeking stability

Many researchers are optimistic about the prospects for seeing elements 119 and 120 in the coming years. But their production rates are likely to be tiny: instead of one or two detections per day, the current techniques might be lucky to spot one a year. So it is a long game, and requires great patience.

However, nuclear scientists have speculated that there exists an "island of stability" for isotopes with "special" numbers of protons and neutrons in their nuclei. Just as electrons in atoms are arranged in shells, so too the protons and neutrons have a shell structure. And just as the shell structure of electrons conveys particular stability to some configurations, especially the fully filled shells of the rare gases, so there are "magic numbers" of protons and neutrons that confer stability. The center of this putative island of stability corresponds to a "doubly magic" nucleus (that is, for both proton and neutron

ABOVE: Kōsuke Morita, the leader of the Japanese team that discovered the new superheavy element 113 (nihonium), points to its position in the periodic table in 2015.

numbers) in the superheavy regime.

The leading candidate for such a doubly magic nucleus is the isotope flerovium-298, with 114 protons and 184 neutrons. If such isotopes do prove to be specially stable, some might have lifetimes long enough for significant amounts of the element to be gradually accumulated. But we just don't know yet if the island of stability awaits out there in superheavy space, and scientists suspect it will be very difficult to reach it any time soon. Will the centuries-long quest to find new elements continue, or are we nearing the end of the road? You can be sure that element-hunters—now turned element-makers—won't slacken their determination to find out.

SOURCES FOR QUOTES

page 11: "We don't hire": Chapman, p.154.

page 14: "The body of": Plato, *Timaeus and Critias*, p.43. Penguin, 1986.

page 16: "Most of the... That from which": Aristotle, *Metaphysics*, Book I, Part 3 (ca. 350 BC).

page 18: "mix into one another": Pullman, p.14.

page 19: "one hundred and sixty-four": J. B. van Helmont, *Oriatrike or Physick Refined*, transl. J. Chandler. Lodowick Loyd, London, 1662.

page 21: "The air round": Aristotle, *Meteorology* Bk I, Pt 3 (ca. 350 BC), transl. E. W. Webster.

page 22: "all things happen": Text designated DK22B80 in the collection of Presocratic sources collected by Hermann Diels & Walther Kranz, *Die Fragmente der Vorsokratiker*. Weidmann, Zurich, 1985.

page 24: "Everything is born": Pullman, p.19.

page 25: "Earth has its place": J. C. Cooper, *Chinese Alchemy*, p.89. Sterling, New York, 1990.

page 27: "We must, of course": Plato, op cit., p.79.

page 28: "The gods used": Plato, ibid, p.78.

page 33: "How innocent, how happy": Multhauf, p.95. "Driven utterly": ibid.

page 45: "came down like a wolf": Lord Byron, "The Destruction of Sennacherib" (1815).

page 45: "The Greek civilization": T. K. Derry & T. I. William, *A Short History of Technology*, p.122. Clarendon Press, Oxford, 1960.

page 46: "Cement steel is nothing": C. S. Smith, "The discovery of carbon in steel", *Technology and Culture* 5, 149–175 (1964), here p.171.

page 52: "could wake up the dead": H. M. Pachter, *Paracelsus: Magic Into Science*, p.137. Henry Schuman, New York, 1951.

page 54: "Tartarean Sulphur": J. Milton, *Paradise Lost*, Bk II, line 69 (1667).

page 61: "like a cannon bullet taken": J. Emsley, *The Shocking History of Phosphorus*, p.32. Macmillan, 2000. "the body of man": ibid, p.34.

page 62: "blood red drops... surpasses the sweetness": L. Thorndike, *A History of Magic and Experimental Science*, Vol. III, p.360. Columbia University Press, New York, 1934.

page 72: "Believe me when I declare": R. Boyle, *The Sceptical Chymist*, p.xiii. London, 1661.

page 73: "Out of some bodies": *The Sceptical Chymist*, in Brock, p.57; "certain primitive": ibid, in H. Boynton, *The Beginnings of Modern Science*, p.254. Walter J. Black, Roslyn, 1948.

page 78: "unknown to the Ancients": Wothers p.32; "was found a metal": A. Barba, *The Art of Metals*, p.89–90. S, Mearne, London, 1674.

page 83: "there is another metal": in Agricola, p.409; "Zink gives the Copper": G. E. Stahl, *Philosophical Principles of Universal Chemistry*, p.335. John Osborn & Thomas Longman, London, 1730; "a great resemblance": Wothers, p.58; "unknown to the European": R. Boyle, *Essays of the strange subtility great efficacy determinate nature of effluviums*, p.19. M. Pitt, London, 1673.

page 84: "has the distinctive": Agricola, p.113.

page 87: "there are also found": Theophilus, *On Diver Arts*, p.59. Dover, New York, 1979.

page 88: "really poisonous... beware": Cennino Cennini, *The Craftsman's Handbook*, transl. D. V. Thompson, p.28 . Dover, New York, 1933.

page 90: "there is no keeping": ibid, p.28.

page 93: "draw anything out": J. B. van Helmont, *Oriatrike, or, Physick Refined*, p.615. Lodowick Loyd, London, 1662; "calx of a new metal": T. Bergman, *Physical and Chemical Essays*, Vol. 2, p.202. J. Murray, London, 1784.

page 98: "As the discovery": M. Klaproth, *Analytical Essays Towards Promoting the Chemical Knowledge of Mineral Substances*, Vol. 1, p.476. T. Cadell, London, 1801.

page 106: "shy and bashful": C. Jungnickel & R. McCorrmach, *Cavendish: The Experimental Life*, p.304. Bucknell, 1999.

page 110: "my breast": J. Priestley, *Experiments and Observations of Different Kinds of Air*. J. Johnson, London, 1775.

page 121: "each of those substances": S. Tennant, "On the nature of the diamond", *Philosophical Transactions of the Royal Society* **87**, 123–127, here p.124 (1797).

page 127: "took advantage of": M. Faraday, "On fluid chlorine", *Philosophical Transactions of the Royal Society* **113**, 160–165, here p.160 (1823).

page 128: "The fire melts": G. Agricola, *De natura fossilium*, transl. M. C. Bandy & J. A. Bandy, p.109, footnote. Mineralogical Society of America, New York, 1955.

page 132: "grey, very hard": "H. V. C. D.", *Journal of Natural Philosophy, Chemistry, and the Arts*, July, pp.145–146 (1798); "On account of": R. Newman, "Chromium oxide greens", in E. West Fitzhugh (ed.), *Artists' Pigments: A Handbook of Their History and Characteristics*, Vol. 3, p.274. National Gallery of Art, Washington DC, 1997.

page 134: "promises to be": F. Stromeyer, "New details respecting cadmium", *Annals of Philosophy* [translated from *Annalen de Physik*], **14**, pp.269–274 (1819).

page 140: "Atoms are round": Brock, p.128.

page 140: "interests of science": J. Dalton, *A New System of Chemical Philosophy*, Preface, v. R. Bickerstaff, London, 1808.

page 147: "small globules... some of which": H. Davy, "The Bakerian Lecture: On some new phenomena of chemical changes produced by electricity, particularly the decomposition of the fixed alkalies...", *Philosophical Transactions of the Royal Society* **98**, 1–44, here p.5 (1808); "bounded about": H. Davy (ed. J. Davy), *The Collected Works of Sir Humphry Davy*, Vol. I, p.109. Smith, Elder & Co., London, 1839–40; "an instantaneous": Davy, "The Bakerian Lecture", p.13.

page 148: "When thrown": Davy, *The Collected Works*, op. cit., p.245.

page 150: "more universally": L. B. Guyton de Morveau, *Method of Chymical Nomenclature*, transl. S. James, p.49. G. Kerasley, London, 1788.

page 155: "candid criticisms": H. Davy, *Elements of Chemical Philosophy*, p.350. J. Johnson & Co., London, 1812; "dark gray": ibid.

page 156: "dark olive coloured": Davy, *Elements of Chemical Philosophy*, p.316; "is more analogous": ibid, p.314.

page 158: "obliged to seek": Davy, *Collected Works*, op. cit., Vol. IV, p.116; "a film of a": ibid, p.120; "a grayish opaque": ibid., p.121; "black particles... numerous gray": ibid., *Elements*, pp.268, 263.

page 159: "there is not the smallest": T. Thomson, *A System of Chemistry*, Vol. I, p.252. Baldwin, Cradock & Joy, London, 1817.

page 163: "It was as though": W. A. Tilden, "Cannizzaro Memorial Lecture", in D. Knight (ed.), *The Development of Chemistry 1798–1914*, 567–584, here p.579. Routledge, London, 1998.

page 164: "I saw in a dream": B. M. Kedrov, "On the Question of the psychology of scientific creativity (on the occasion of the discovery of D. I. Mendeleev of the periodic law)", *Soviet Psychology* **5**, 18–37 (1966–67).

page 170: "nothing but a pulse": T. Birch, *The History of the Royal Society*, Vol. 3, 10–15, here p.10 (1757); "the vast interplanetary": W. D. Niven (ed.), *The Scientific Papers of James Clerk Maxwell*, Vol. 2, LIV, pp.311–323, here p.322. Cambridge University Press, 1890.

page 171: "telegraphy without wires": W. Crookes, "Some possibilities of electricity", *Fortnightly Review* **51**, 175 (1892).

page 173: "two splendid blue... The bright blue light": G. Kirchhoff & R. Bunsen, "Chemical analysis by spectrum-observations", Second Memoir, *The London, Edinburgh, and Dublin Philosophical Magazine and Journal of Science*, **22,** p.330. 1861.

page 176: "waiting to be... I have seen": W. H. Brock, *William Crookes (1832–1919) and the Commercialization of Science*, p.63. Ashgate, Aldershot, 2008.

page 177: "the green line": W. Crookes, "Further remarks on the supposed new metalloid", *The Chemical News* **3(76)**, p.303 (1861).

page 181: "I... began to": M. W. Travers, *A Life of Sir William Ramsay*, p.145. Edward Arnold, London, 1956.

page 182: "but it appears": W. Ramsay, *The Gases of the Atmosphere: The History of Their Discovery*, p.195. Macmillan, London, 1915.

page 183: "the presence of... combine with argon": H. G. Wells, *The War of the Worlds*, in H. G. Wells, *The Science Fiction*, Vol. I, p.317. J. M Dent, London, 1995.

page 184: "a blaze of crimson": M. W. Travers, *The Discovery of the Rare Gases*, pp.95–6. Edward Arnold, London, 1928.

page 186: "The question was": C. Nelson, *The Age of Radiance: The Epic Rise and Dramatic Fall of the Atomic Era*, p.25. Scribner, New York, 2014.

page 187: "I had a passionate": R. W. Reid, *Marie Curie*, p.65. Collins, London, 1974.

page 188: "a metal never before": S. Quinn, *Marie Curie: A Life*. Da Capo Press, 1996; "we had an especial joy": M. Curie, *Pierre Curie*, p.49. Dover, New York, 1963; "One of our joys": ibid, p.92.

page 208: "to change": R. Rhodes, *The Making of the Atomic Bomb*, p.140. Simon & Schuster, New York, 1986.

FURTHER READING

G. Agricola, *De Re Metallica*, transl. H. C. Hoover & L. H. Hoover. Dover, 1950.

H. Aldersey-Williams, *Periodic Tales*. Penguin, 2011.

P. Ball, *The Elements: A Very Short Introduction*. Oxford University Press, 2004.

W. H. Brock, *The Fontana History of Chemistry*. Fontana, 1992.

K. Chapman, *Superheavy: Making and Breaking the Periodic Table*. Bloomsbury, 2019.

J. Emsley, *Nature's Building Blocks*. Oxford University Press, 2001.

M. D. Gordin, *A Well-Ordered Thing: Dmitrii Mendeleev and the Shadow of the Periodic Table*. Basic Books, 2004.

T. Gray, *The Elements*. Black Dog, 2009.

R. Mileham, *Cracking the Elements*. Cassell, 2018.

R. P. Multhauf, *The Origins of Chemistry*. Gordon & Breach, 1993.

B. Pullman, *The Atom in the History of Human Thought*. Oxford University Press, 1998.

E. Scerri, *The Periodic Table: Its Story and Its Significance*, 2nd edn. Oxford University Press, 2020.

E. Scerri, *The Periodic Table: A Very Short Introduction*. Oxford University Press, 2019.

E. Scerri, *A Tale of Seven Elements*. Oxford University Press, 2013.

P. Wothers, *Antimony, Gold, and Jupiter's Wolf*. Oxford University Press, 2019.

Picture credits

Every attempt has been made to trace the copyright holders of the works reproduced, and the publishers regret any unwitting oversights. Illustrations on the following pages are generously provided courtesy of their owners, their licensors, or the holding institutions as below:

Alamy Stock Photo: 16 (The Picture Art Collection); 21 (Granger Historical Picture Library); 35 top (Album); 44 (www.BibleLandPictures.com); 45 (FLHC 40); 55 top (Laing Art Gallery, Newcastle-upon-Tyne/Album); 55 bottom (Biblioteca Medicea Laurenziana, Florence); 56 (CPA Media Pte Ltd); 70–71 (Science History Archive); 76 top (Interphoto); 125 (Institution of Mechanical Engineers/Universal Images Group North America LLC)

© akg-images: 42 (SMB, Antikenmuseum, Berlin/ Bildarchiv Steffens); 87 (Topkapi Museum, Istanbul/Roland and Sabrina Michaud); 74–75 (Annaberg, Sachsen, Stadtkirche St. Annen)

Annaberg-Buchholz (St Ann's Church) / Wikimedia Commons (PDM): 96–97

Bayerische Staatsbibliothek München, Chem. 118 d-1, p.457 (detail): 162 bottom

Bethseda, The National Library of Medicine: 188 bottom

Biblioteca Civica Hortis, Trieste (PDM): 57 top

Biblioteca General de la Universidad de Sevilla (CC 1.0): 47

Bibliothèque nationale de France, département Estampes et photographie: 157 top

The British Library, London (PDM 1.0): 14, 53

Cavendish Laboratory, University of Cambridge, after J.B. Birks, ed., Rutherford at Manchester (London: Heywood & Co., 1962) p.70: 195

Deutsches Museum, Munich: 178

Edgar Fahs Smith Collection, Kislak Center for Special Collections, Rare Books and Manuscripts, University of Pennsylvania: 162 top

© E. Galili: 40

© Ethnologisches Museum der Staatlichen Museen zu Berlin—Preußischer Kulturbesitz (bpk); Photo: Ines Seibt: 33

Finnish Heritage Agency - Musketti (CC by 4.0): 139

Francis A. Countway Library of Medicine, Harvard: 129, 186

Gerstein—University of Toronto: 79, 120

Getty Images: 18 (© DEA / G. Nimatallah/De Agostini); 59 (Derby Museum and Art Gallery); 99, 133 (Fine Art Images/Heritage Images); 149 (Hulton Archive); 160–161 (Bettmann); 164 bottom, 184 bottom; 190–191 (Corbis); 193 top; 194 (© Science Museum /SSPL); 217 (Kazuhiro Nogi/Afp)

Getty Research Institute, Los Angeles: 15, 23, 41, 63, 64 (right); 80

GSI Helmholtzzentrum für Schwerionenforschung GmbH; photo: A. Zschau: 215

© Gun Powder Ma/Wikimedia Commons (CC0 1.0): 43

© History of Science Museum, University of Oxford: 11

Homer Laughlin China Company, 'Fiesta' is a registered trademark of the Fiesta Tableware Company; Photo: courtesy Mark Gonzalez: 100

© Institute of Nautical Archaeology, Texas: 86 bottom

J. Paul Getty Museum, Villa Collection, Malibu, California: 12–13, 35 bottom

Library of Congress, Washington, DC: 49, 119, 185 (Prints and Photographs Division); 107 (Rare Book and Special Collections Division)

The Linda Hall Library of Science, Engineering & Technology, courtesy of: 130, 201

Manna Nader, Gabana Studios, Cairo, by kind permission: 30–31

Marco Bertilorenzi, after 'From Patents to Industry. Paul Héroult and International Patents Strategies,1886-1889' (2012): 159 right

The Metropolitan Museum of Art, New York: 34, 38, 46 bottom, 90, 111

© Michel Royon / Wikimedia Commons (CC BY 2.0): 153 top

Ministry of Tourism and Antiquities, Cairo, courtesy Egymonuments.gov.eg; Photo: Ahmed Romeih - MoTA: 153 bottom

Musée Curie (Coll. ACJC), Paris: 188 top

Naples, by permission of the Ministry for Cultural Heritage and Activities and for Tourism— National Archaeological Museum; photo: Luigi Spina, inv. 5623: 20

National Central Library of Florence: 117

National Galleries of Scotland: 114

National Gallery of Art, Washington, DC. Samuel H. Kress Collection: 87

National Gallery, London: 134

Natural History Museum Library, London: 123, 179, 180

National Library of Norway, via Project Runeberg, DRM Free: 86 top

National Museum of China, Beijing/photo: BabelStone//Wikimedia Commons (CC BY-SA 3.0): 57 bottom

National Portrait Gallery, Mariefred, Södermanland: 84

Oak Ridge National Laboratory, PDM: 205

Philip Stewart, 2004, by kind permission: 166–167

© The President and Fellows of St John's College, Oxford: 21

Qatar National Library: 126

Rawpixel Ltd (CC0), via Flickr: 168–169

© 2010-2019 The Regents of the University of California, Lawrence Berkeley National Laboratory: 198, 200, 211 (Photo: Donald Cooksey); 199, 120 (Photo: Marilee B. Bailey);

204, 206–207 (Photolab); 197 (Photo: Roy Kaltschmidt).

The National Library of Scotland, Reproduced with the permission of: 155

The Royal Library, Stockholm, Archive of Swedish cultural commons: 138

The Royal Society, London: 56 bottom (CLP/11i/21 - detail), 127

Schlatt, Eisenbibliothek: 89

Science & Society Picture Library—All rights reserved: 141 (© Museum of Science & Industry); 165, 177 (Science Museum, London)

Science History Institute, Philadelphia (PDM 1.0), courtesy of: 48 top, 72, 64 left, 65, 72, 93, 95, 105, 109, 116 (Douglas A. Lockard), 124, 135, 163, 172, 173, 175, 181

Science Photo Library, London: 121; 148 (Royal Institution Of Great Britain); 150–151 (Sheila Terry); 174 (Rhys Lewis, Ahs, Decd, Unisa / Animate4.Com)

SLUB Dresden / Deutsche Fotothek: 50–51

Smithsonian Libraries, Washington, courtesy of, via BHL: 27, 76 bottom, 92, 154

Stadtarchiv Wesel, O1a, 5-14-5-02: 196

Stanford Libraries, courtesy, David Rumsey Map Center (PURL https://purl.stanford.edu/ bf391qw5147): 26 (The Robert Gordon Map Collection), 28 (The Barry Lawrence Ruderman Map Collection); 29 (Glen McLaughlin Map Collection of California)

Statens Museum for Kunst, Copenhagen: 158

© Szalax / Wikimedia Commons (CC by 4.0): 37

Tekniska museet, Photo: Lennart Halling, 1960-11: 137

Topkapi Sarayi Ahmet III Library, Istanbul/ Wikimedia Commons (CC BY 4.0): 7

© The Trustees of the British Museum: 39, 82

University of California Libraries, via BHL: 183, 193 bottom

University of Illinois Urbana-Champaign, via BHL: 17

University of Miskolc: 66

United States Department of Energy, Office of Public Affairs, Washington: 198, 209; 203 (Special Engineering Detachment, Manhattan Project, Los Alamos, Photo: Jack Aeby)

United States Patent and Trademark Office, www.uspto.gov: 48 bottom, 159 left

Vassil/Wikimedia Commons (PDM): 85

Victorian Web, photo: Simon Cooke: 184 top

The Walters Art Museum, Baltimore (CC0): 24, 62

Wellcome Collection, London (CC BY 4.0): endpapers, 25, 67 both, 68, 72, 77, 81, 83, 91, 101, 102–103,108, 112, 115, 128, 132, 140, 142–143, 144, 145, 146, 147, 152, 164 top, 170, 176, 182, 187 both, 189

Wikimedia Commons (PDM): 122

Yoshida-South Library, Kyoto University: 10

INDEX

28 janv — tube capillaire, 1 mètre long
2 jours — ? 0,7 mm diamètre intérieur

en ouvrant
peu odeur ozone

app 0... — violent
½ heure après apr (6h/3ᶜ) — 2000 — 18ˢ

parafine
aluminium
taile métall
métal
galaz
verre

2 jours (75)
agi[ssa]nt à travers
aluminium
rien

6ʰ avec air à travers galaz
2000 — 3ˢ (app 0)
le même que pour vide mais
galaz changé aspect

galaz

6ʰ avec air à travers paraf
de même que pour vide
2000 — 3ˢ apr. 0